SCIENCE IN MOSCOW

MEMORIALS OF A RESEARCH EMPIRE

SCIENCE IN MOSCOW

MEMORIALS OF A RESEARCH EMPIRE

ISTVÁN HARGITTAI

MAGDOLNA HARGITTAI

World Scientific

NEW JERSEY • LONDON • SINGAPORE • BEIJING • SHANGHAI • HONG KONG • TAIPEI • CHENNAI • TOKYO

Published by

World Scientific Publishing Co. Pte. Ltd.
5 Toh Tuck Link, Singapore 596224
USA office: 27 Warren Street, Suite 401-402, Hackensack, NJ 07601
UK office: 57 Shelton Street, Covent Garden, London WC2H 9HE

British Library Cataloguing-in-Publication Data
A catalogue record for this book is available from the British Library.

This edition in English is based on the book in Hungarian, *Moszkvai séták a tudomány körül* (Budapest: Akadémiai Kiadó, 2018).

The images on the front cover were purchased by the Publisher from an outside source (Shutterstock)

SCIENCE IN MOSCOW
Memorials of a Research Empire

ISBN 978-981-120-344-2

Typeset by Stallion Press
Email: enquiries@stallionpress.com

Printed in Singapore

Also by the Authors

Culture and Art of Scientific Discoveries: A Selection of Istvan Hargittai's Writings (Balazs Hargittai, Ed., Springer 2019)

Istvan Hargittai and Magdolna Hargittai, *New York Scientific: A Culture of Inquiry, Knowledge, and Learning* (Oxford University Press, 2017)

Balazs Hargittai and Istvan Hargittai, *Wisdom of the Martians of Science: In Their Own Words with Commentaries* (World Scientific, 2016)

Magdolna Hargittai, *Women Scientists: Reflections, Challenges, and Breaking Boundaries* (Oxford University Press, 2015)

Istvan Hargittai and Magdolna Hargittai, *Budapest Scientific: A Guidebook* (Oxford University Press, 2015)

Balazs Hargittai, Magdolna Hargittai, and Istvan Hargittai, *Great Minds: Reflections of 111 Top Scientists* (Oxford University Press, 2014)

Istvan Hargittai, *Buried Glory: Portraits of Soviet Scientists* (Oxford University Press, 2013)

Istvan Hargittai, *Drive and Curiosity: What Fuels the Passion for Science* (Prometheus, 2011)

Istvan Hargittai, *Judging Edward Teller: A Closer Look at One of the Most Influential Scientists of the Twentieth Century* (Prometheus, 2010)

Magdolna Hargittai and Istvan Hargittai, *Symmetry through the Eyes of a Chemist,* 3rd Edition (Springer, 2009; 2010)

Magdolna Hargittai and Istvan Hargittai, *Visual Symmetry* (World Scientific, 2009)

Istvan Hargittai, *The DNA Doctor: Candid Conversations with James D. Watson* (World Scientific, 2007)

Istvan Hargittai, *The Martians of Science: Five Physicists Who Changed the Twentieth Century* (Oxford University Press, 2006; 2008)

Istvan Hargittai, *Our Lives: Encounters of a Scientist* (Akadémiai Kiadó, 2004)

Istvan Hargittai, *The Road to Stockholm: Nobel Prizes, Science, and Scientists* (Oxford University Press, 2002; 2003)

Balazs Hargittai, Istvan Hargittai, and Magdolna Hargittai, *Candid Science I–VI: Conversations with Famous Scientists* (Imperial College Press, 2000–2006)

Istvan Hargittai and Magdolna Hargittai, *In Our Own Image: Personal Symmetry in Discovery* (Plenum/Kluwer, 2000; Springer, 2012)
Istvan Hargittai and Magdolna Hargittai, *Symmetry: A Unifying Concept* (Shelter Publications, 1994)
Ronald J. Gillespie and Istvan Hargittai, *The VSEPR Model of Molecular Geometry* (Allyn & Bacon, 1991; Dover Publications, 2012)

Preface

This book introduces the reader to the visible reminders of science and scientists in Moscow. We use the term *science* more broadly than generally understood in the English language; we take it more closely to the all-embracing Russian word of science, *nauka*. In addition to the natural sciences, we include medicine, technologies, and agricultural sciences as well.

In Moscow, there is an extraordinary number of statues, busts, and memorial plaques devoted to scientists, rivaled only by the memorials of politicians and military leaders. This stems from the Russian reverence toward tradition, history, and authority, and from the great concentration of scientific and higher educational institutions in Moscow. It is very probable that Moscow has more memorials of scientists than any other city in the world. This city with its surroundings is comparable to a medium-size European country; Moscow proper has about 12 million people with millions more in its surroundings. It is the capital of a highly centralized country of over 140 million people and used to be the capital of the highly centralized much larger Soviet empire.

There were milestone achievements of Russian and Soviet science during the last decades of the nineteenth century and in the twentieth century. Under the Soviets, some areas of science — certainly not all — thrived. The overall poor infrastructure was countered by focusing resources on a few privileged areas of research and development.

As the life of society in general, that of the scientific community was also rife with contradictions. Even the families of many of the topmost scientists and many of these scientists themselves suffered from the paranoiac persecution of perceived enemies of the people by the Soviet authorities. The later head of the space program, Sergei Korolev, toiled in a slave labor camp for years. Lev Landau spent a year incarcerated in the infamous Lubyanka only to become one of the key theoretical physicists of the nuclear weapons program; then, to win the Nobel Prize. Under Stalin, family members of at least two future Nobel laureates and the brothers of two future presidents of the Soviet Academy of Sciences perished.

On the other hand, the perils facing the Soviet Fatherland in World War II and the real or perceived dangers in the Cold War greatly strengthened devotion to the Soviet order. Even some of the most apolitical scientists joined the Communist Party during the fateful years of WWII, which in Russian speech is labeled as the Great Patriotic War.

In the post-war period, up to the collapse of the Soviet Union, science offered the most attractive career for gifted young people. It was a career where there was still some room for individual initiative and competition that true talent craved. It gave perks and privileges, to some even foreign travel, in this hermetically closed country. Other attractive careers that rival science for gifted young people in the West, such as business, advertising, finance, jurisprudence, and politics, did not exist or did not represent credible alternatives.

Erecting monuments for scientists was a sign of recognition of greatness in most instances. Then, there have been additional reasons for such commemoration. For example, some insecurity and a need for laying claim to priorities vis-à-vis the West. In discussing the achievements of Russian/Soviet science in Moscow, one often hears complaints about the lack of proper recognition. The Soviet system itself contributed much to such a state of affairs. On the one hand, it kept some of its greatest scientists and technologists in the dark for real or perceived security reasons. Suffice it to mention again Sergei Korolev who was unknown not only to the world but at home as well. Upon his death, he became an instant celebrity and memorials sprung up honoring his success in the Soviet conquest of space. On the other hand, often Soviet science historiography ascribed priorities in various discoveries to Russian and Soviet scientists when it was not justified. This eroded the credibility of valid claims.

The vast majority of the memorials of scientists were erected by official initiative and using state funds. This introduced a great amount of political bias into who received recognition and who did not. Still most, even if not all, of those scientists who are thus remembered made considerable contributions to science. However, many deserving scientists of comparable achievements did not get such remembrance because for some reason they had fallen out of favor or had never been in favor with the authorities. A few prominent martyrs of repression have received commemoration following the collapse of the Soviet order, but numerous others did not. Even when martyr scientists are commemorated, their tragic fates may be masked implying that facing the past was too painful and irreverent with respect to national prestige. There are then, of course, those scientist martyrs that could not fulfill the promise of their talent and ambition because they had been destroyed at an early stage of their career. We need to keep in mind such facets of history while justifiably admiring the existing arsenal of memorials devoted to scientists.

The memorials of scientists — as those of the politicians and military leaders — help to conserve the Soviet-era values and hierarchies. This has happened not only in such visible aspects in the life of Russian society as the monuments. Let us mention here a non-conspicuous, yet telling example. We describe what happened in an admittedly simplified way. People used to live in state-owned apartments in Moscow. When the political changes came these apartments were privatized, and the inhabitants became the owners overnight. This was no mere gift because with the ownership came the obligations of maintenance and renovation. Still, this approach meant that the results of distribution of the quality and quantity of living space were frozen according to what had ensued under the Soviet system. Those who had been accorded spacious apartments in quality neighborhoods became owners of spacious apartments in those neighborhoods. Those who had a tiny room for a family in a communal apartment were left with such conditions. Those who had had nothing became the owners of nothing. This solution was simple and efficient, but conserved Soviet injustices.

The memorials that had been erected under the Soviet regime to honor scientists have never been revised so the injustices live on. There have been very few exceptions. There were some sculptures of Trofim Lysenko and Iosif Stalin sitting together and those memorials have disappeared. In the subsequent pages there will be a few statues that commemorate scientists whose career had been helped by their accusations of others who then perished. Even for the memorials of politicians no consistent revision has taken place. Vladimir Lenin's mummified body is still on display in the Lenin Mausoleum on the Red Square. The queues of those waiting to see it are shorter than they used to be, however, Lenin still holds a commanding presence in Moscow with monumental statues. Even Stalin has a bust on the Red Square, which is never without fresh flowers. Ever since the first attempts to unmask Stalin's crimes, there has been hesitation of varying degrees in how officialdom should be relating to him and his legacy. A poster in 2016 paying respect to the greatest military leaders in Russian history in the foursome of Peter the Great, Suvorov, Kutuzov, and Stalin, does not raise an eyebrow.

Statues of Nobel Peace Prize laureate human rights activist physicist Andrei Sakharov (forefront) and the founder of the infamous secret police Feliks Dzerzhinsky in the Muzeon Park.

There is a sculpture garden, the Muzeon (see, Chapter 8), beautifully located in the heart of Moscow, where many artifacts of the Soviet regime have been collected. Perhaps, their artistic and historical values warrant preservation. They include a number of sculptures of past political leaders, such as Stalin, Feliks Dzerzhinsky, and Leonid Brezhnev. There are also moving memorials of Stalin's victims — many of whom were scientists — in the same sculpture garden. The impressive Stalin statue with his partly disfigured face stands on the background of masses of unrecognizable heads of his victims. In another spot, an unassuming statue of the Nobel Peace Prize laureate human rights advocate physicist, Andrei Sakharov, contrasts the nearby monumental statue of the founder of the secret police Dzerzhinsky. Such juxtapositions are powerful statements.

Many of the memorials of scientists in Moscow reflect value judgment from Soviet times. We stress such a caveat for proper perspective. Without keeping this caveat in mind we might inadvertently add to the frequent falsification of history. The development of great science in the Soviet Union had both uplifting and tragic aspects. The opening of higher education for broad masses, following the Bolshevik revolution, brought out talent that might have not surfaced otherwise. The new regime freed unprecedented energies among layers of society that had been left out of the possibility of rising to the top in the professions. At the same time, the Soviet takeover led to systematic efforts of replacing the old intelligentsia with a new one. This seldom happened in the gradual way by cultivating a new one and waiting for the old to disappear, the changes taking their natural course. Rather, mostly it was done by force. Stalin's famous first terror called sometimes the *Great Purge*, happened in the mid-1930s. However, the liquidation of the old intelligentsia had begun already in the 1920s. The processes were not sterile because some members of the old were spared, even incorporated into the new and sent to the West to rebuild interactions.

Some branches of science and their scientists commanded great respect in the Soviet Union and were accorded exceptional treatment. Those who were expected to develop the atomic bombs and then the hydrogen bombs; who built rockets with warheads that could reach the enemy at any distance on the planet; who created sputniks and spaceships, were the privileged ones. They were showered with *dachas* (weekend homes in the suburbs); comfortable apartments, cars, and free travel within the country; education of their children in a university of their choice; and academy membership and scientific degrees. Academy membership implied further perks and privileges.

The Academy of Sciences had an exceptionally high prestige in the Soviet Union. Many of the members of the Academy were scientists of high quality and deserved this distinction. There were then those that the Party wanted to be in the Academy for political considerations. And there were those for whom it was recognition of their exceptional contribution to defense without it representing a scientifically exceptional achievement. Conversely, many significant scientists were never elected to the Academy. In the 1930s and 1940s, many of the scientists who fell out of favor perished. Later, dissenting scientists may have lost their jobs and livelihoods; others may have just not received support for their research, or may have just been left without recognition of their achievements. Thus, under these conditions where memory honored by a statue, a bust or a memorial plaque appears to be arbitrary, we must take into account the historical circumstances or else we enhance the injustices.

Our *Science in Moscow* follows two previous books, *Budapest Scientific* and *New York Scientific*. We have stated in our books about Budapest and New York that no science history should be compiled on the bases of the memorials. This caveat is even more valid for Moscow. However, having issued the caveats above, the amount and quality of these memorials deserve our recognition and there is much to learn about the scientists and their discoveries represented by them. There is also much to learn about Soviet and Russian history by observing who is represented and who is left out and in what ways they are remembered and why this has happened.

We mentioned our personal interest in the case of our books of the scientific memorials in Budapest and New York, and there was certainly a special bond in the case of Moscow as well. We both have been in Moscow for visits, had run experiments there for our research, speak Russian, and have scores of friends. One of us (IH) was a Master's degree student in chemistry at the Lomonosov Moscow State University in 1961–1965 and has the best of memories of the professors and fellow students. The early 1960s carried some promise for improvement in the dire Soviet conditions, and although the promise never developed into fulfillment, it lent some optimism while it lasted.

In our global scientific and human interactions, we have experienced considerable barriers between the East and the West, much of which have remained. In the 1970s and 1980s, some referred to our interactions as building a bridge. We like to think that this book may be another brick in that bridge.

Chapter Overviews

1 *Academy of Sciences*

The Academy is comprised of a body of leading scientists — the members — and a network of research institutes. The old headquarters of the Academy, now the seat of its Presidium, is the former palace of a wealthy Russian family; the new headquarters is a modern building whose impressive appearance contrasts the diminishing role of this great institution. A considerable portion of the vast nationwide network of research institutes is situated around the eastern side of the wide Leninsky Avenue. Some of the buildings of these institutes were built in the 1950s in lavish old-time elegance. The institutes carry the names of their founding scientists and display busts and memorial plaques paying tribute to them.

Seven of the physicists connected with the Physical Institute received the Nobel Prize. That one of them, Andrei Sakharov was a Nobel Peace Prize laureate only underscores the overall impact of physicists in Soviet society.

Igor Kurchatov was forty years old when Stalin appointed him nuclear tsar. He worked with the best Soviet physicists and benefited from first-class intelligence about the American nuclear weapons program. He loomed larger than life over the nuclear empire that was exempt from ideological purity and the anti-Semitic hiring restrictions practiced nationwide in elite institutions.

Nikolai Vavilov was a world-renowned plant geneticist who first supported and then tried to unmask the charlatan Trofim Lysenko's unscientific teachings. Stalin thought Lysenko's approach applicable to transforming human society and Vavilov had to perish. The next Soviet leader, Nikita Khrushchev, was also taken by Lysenko. That the institute of genetics that was once under Lysenko's directorship today carries Vavilov's name illustrates the complexity of the story of biological sciences in Russia.

2 *Moscow University*

This flagship institution of Soviet/Russian higher education was named after the polymath Mikhail Lomonosov who had himself been labeled "the first Russian university." The University used to be in downtown Moscow without much space for expansion. In the early 1950s, seven new high-rises were being built in Moscow, and Stalin was looking for worthy occupants. The University received the one, of the seven, with the best physical location and virtually unlimited space for expansion. Memorials of scientists abound in the walkways and plazas and in the interior of the iconic main building.

3 *Earth Science Museum*

The Earth Science Museum at Moscow University has the character of a natural history museum. It is geared to assist education of students of this institution and other students that come to utilize its exceptional collection and environment. A special feature of the collection is the large number of busts honoring scientists, especially earth scientists and explorers, which in every turn stress the human contribution to our knowledge. There was a special program of creating these busts by outstanding sculptors in the early 1950s when the country's flagship university was preparing to move into its new venue.

4 *Technology and Technologists*

Vladimir Shukhov epitomized the talent and inventiveness of the best of engineering and he is remembered by his landmark tower. The number of engineers in the Soviet Union and in Russia has been far larger than any other profession belonging to the subject matter of our book. The engineers and technologists contributed much to the Soviet/Russian success stories from exploring the vital oil and gas reserves to pioneering space research. The large masses of engineers received over-specialized training, but the top layer of technologists deserved the recognition the state afforded them. They formed part of the elite whose memories are kept alive in numerous memorials. Oil and gas have financed much of the economy whereas the space program added more than anything else to the prestige of the nation — in fact there was hardly anything else to be proud of. Conquering space was a derivative of the military preparedness in designing long range rockets and missiles. The most conspicuous memorials in Moscow were erected to honor the Soviet conquerors of space.

5 *Medical City*

Within a short walk from the iconic Novodeviche Cemetery, there is the campus of the Sechenov Medical University with quite a few of the vast conglomerate of health-related institutions spread all over Moscow. The large number of memorials to physicians, professors of medicine, and innovators in the health industry is heartwarming. This is in contrast with the fact that under the Soviet system, health-related activities were not kept in especially high esteem. They were not considered to belong to the means of production, like heavy industry, for example. On the background of neglect and isolation, there were spikes of extraordinary accomplishments in medicine.

6 *Timiryazev Academy*

Few areas suffered from pseudoscience and mismanagement as did Soviet agriculture. There were outstanding early scientists like the plant physiologist Kliment Timiryazev and the plant geneticist Nikolai Vavilov. Timiryazev died just as the Soviets were taking over and became an icon. For many others who refused to fall in with the party line, the executioner's bullets ended their lives. Bringing back the science of biology and agriculture from demise has been a heroic act and the only hope Russian agriculture can have for its future.

7 *Novodeviche Cemetery*

This cemetery is a concentrate of reverence to past glory and authority. There are thousands of graves of revolutionaries, political leaders, military commanders, artists, actors and actresses, poets, writers, singers, and other members of the Russian and Soviet elite, including hundreds of scientists, technologists, explorers, and physicians. It is a sculpture garden, because many of the tombstones are statues, busts, and reliefs. Getting buried in the Novodeviche meant an exceptionally high status in Soviet society; it was like belonging to socialist nobility. The comparison is apt because close family members gained the right to be buried there as well.

8 *Keep Walking*
There are Sakharov memorial places; artistic creations in museums and artists' workshops; the Muzeon sculpture garden; the Darwin Museum and other institutions presenting natural history; memorials in subway stations; the statue of the printer Fedorov and the sculpture of the Leichoudes brothers; memorial plaques on famous buildings, such as the House on the Embankment and 9 Tverskaya Street; other burial places apart from the Novodeviche, such as the Kremlin Wall and a sampler of high-prestige cemeteries; and other venues. Our book is an open-ended project calling for our readers to carry it on.

*****

We have ourselves taken the vast majority of the photographs shown in this book and the source is not indicated for them. Most other images come from 23 Russian friends. They kindly augmented our collection of images and their names are recorded in the captions of the images and in the Acknowledgments. A few additional images came from other sources and we documented them to the best of our knowledge and where necessary we collected the permissions to use them. In cases where readers have further information, we would appreciate hearing about it to augment our efforts.

Acknowledgments

We thank the Hungarian Academy of Sciences for support and Moscow State University for hospitality.

We thank the following for their most diverse assistance: Ekaterina Altova (MSU), Boris Altshuler (FIAN), Nikolai Andreev (Steklov Institute), Tatyana Avrutskaya (NI Vavilov Institute), Tatyana Balakhovskaya (Kapitsa Institute), Ekaterina Bartashevich (Chelyabinsk), Giovanni Battimelli (Università Sapienza, Roma), Galina Bous (Saratov), Nikolai Dolbilin (Steklov Institute), Victoria Dorman (Princeton, Vitaly Ginzburg's granddaughter), Olga Dorofeeva (MSU), Evgeny Dubinin (Earth Science Museum MSU), Behrooz and Mehdi Esrafili (Tehran), German Fadeev (Bauman University), Ekaterina Fadeeva (MSU), Anatoly Fomenko (MSU), Ludmila Frolova (Engelhardt Institute), Yury Gerasimov (Bauman University), Igor and Katya Gomberg (West Stockbridge, MA), Yulia Gorbunova (Kurnakov Institute), the late Boris Gorobets (geologist and writer; Lev Landau's biographer), Anatoly Ishchenko (Lomonosov University of Chemical Technology), Nadezhda Ismailova (Blagonravov Institute), Jan Kandror (Wiesbaden, Germany), Alexei Khokhlov (MSU), Darya Kobyatskaya (Shumakov Institute), Evgeny Koshkin (Timiryazev Academy), Bela Koval (Sakharov Archives), Valery Lunin (MSU), Elena Lvova (MSU), Artemy Maslov (Bauman University), Lyubov Mikhailova (Kerldysh Institute), David Nikogosyan (Cork, Ireland), Marina Ovchinnikova (Burdenko Military Hospital), Sergey Petukhov (Blagonravov Institute), Sofia Prokofieva (Moscow), Vasily Ptushenko (Semenov Institute), Dmitry Pushcharovsky (MSU), Inga Ronova (INEOS), Viktor Sadovnichy (MSU), Alexei Semenov (MSU; grandson of Nikolai Semenov and Yuly Khariton), Igor Shishkov (MSU), the late David Shoenberg (Cambridge, UK), Simon Shnoll (Pushchino), Maya Sirotina (Mendeleev University), Konstantin Skripenko (Earth Science Museum MSU), Julia and Keld Smedegaard (London), Vladimir Tsirelson (Mendeleev University), the late Boris Vainshtein (Shubnikov Institute), Alexander Varshavsky (Pasadena, CA), Stanislav Velichko (Timiryazev Academy), Aleksandr Vernyi (Kurchatov Institute), Anna Vilkova (MSU), the late Lev Vilkov (MSU), Oleg Voloshin (Institute of Biomedical Problems), Larissa Zasourskaya (MSU), and Natalya Zavoiskaya (Moscow).

Special thanks are due to Olga Dorofeeva for her multifaceted assistance following decades of our scientific interactions and to those who read one or a few chapters in draft and made valuable comments: Aleksei Semenov, Stanislav Velichko, and Aleksandr Verny; and to Irwin Weintraub and Robert Weintraub of Beer Sheva who read all chapters and made many valuable comments and suggestions. We much appreciate the efforts of desk editor Ms. Sandhya Devi and other associates of the Publisher in bringing out this attractive volume. This is yet another project in our close to three decades of enjoyable and fruitful cooperation with World Scientific.

Contents

Also by the Authors		v
Preface		vii
Acknowledgments		xvii
Chapter 1	Academy of Sciences	3
Chapter 2	Lomonosov University	67
Chapter 3	Earth Science Museum	101
Chapter 4	Technology and Technologists	131
Chapter 5	Medical City	179
Chapter 6	Timiryazev Academy	225
Chapter 7	Novodeviche Cemetery	255
Chapter 8	Keep Walking	305
Select Bibliography		351
Index		359
Index of Artists and Architects		381

We dedicate this book to the memory of
scientist victims of oppressive regimes

Headquarters of the Russian Academy of Sciences (RAS) and of the Federal Agency of Scientific Organizations (FASO), nicknamed as "the golden brains," 32a Leninsky Avenue, as seen from the balcony of the 32nd floor of the Moscow University tower. The small towers of the main building of the Semenov Institute of Chemical Physics are visible in the forefront and the Shukhov Tower appears behind the Academy headquarters, slightly to the left.

1

Academy of Sciences

There was the Imperial Russian Academy of Sciences, then the Soviet Academy of Sciences, and now, again, the Russian Academy of Sciences. There used to be also the Academy of Medical Sciences and the Academy of Agricultural Sciences. There was some tentative hierarchy between them in that membership in the Academy of Sciences was considered to be higher than membership in either of the other two academies.

The building of the Presidium of the Russian Academy of Sciences, 14 Leninsky Avenue.

A new federal law in 2013 united the three academies under the Russian Academy of Sciences (RAS). This change — along with the academy properties having become state properties — signaled a diminishing role of science in the new Russian order. The Academy of Sciences is a state institution. There are still three additional separate academies, those of education, of architecture and construction sciences, and of the arts — all state institutions. In addition, there are countless academies — non-state institutions — that enjoy much less prestige than the state institutions.

The 2013 law created a Federal Agency of Scientific Organizations (FASO). It was also determined that FASO will be in charge of all properties of the expanded RAS. In 2014, the Russian government approved the list of those 1,007 (!) institutions that were placed under the jurisdiction of FASO. These institutions used to belong to the three previously independent academies.

In the Soviet framework, in addition to the Soviet Academy of Sciences, 14 of the 15 republics each had its own academy of sciences. The Russian Soviet Federal Socialist Republic was the only exception, which was a tacit recognition that Russian scientists dominated the Soviet Academy.

The Russian Academy of Sciences is a vast organization and there is no attempt here to present its history, accomplishments, and institutions. Our aim is to single out a few institutes and introduce the reader to memorials that tell us something about their scientists. All this is to a great extent refers to Soviet times. The Academy of Sciences was an integral part of the Soviet system. The academy membership was probably the most coveted recognition, and the grip of the Communist Party on its activities was tight.

At the same time, there was probably no other institution in the Soviet Union that could have produced such early cracks in the Soviet system as the members of the Academy. Just think of the initiatives of the physicists in 1958 when they wanted the President of the Academy to give an account of his activities before he might be re-elected for a second term. Of all the Soviet institutions and organizations, the Science Academy was unique in preserving some features of democracy in its operations. The intentions and instructions of the government and the Communist Party were not always fully complied with in the elections of the new members. In the post-Stalin period, its members could not be excluded from the Academy even if they had lost employment. This meant that they continued receiving their substantial remuneration as members of the Academy. The human rights activist Andrei Sakharov was an example; this was his source of income during his long years of exile between 1980 and 1986. The current political leadership of Russia has introduced drastic changes in the status and ownership conditions of the Academy that curtailed its independence. This happened under circumstances that are supposed to be much freer than they were in Soviet times. Yet there was hardly any protest or dissent strong enough that would have made the political leadership blink. The drastic changes happened without consultations with the interested parties and virtually overnight.

The Academy of Sciences was created under the auspices of Peter the Great. Stalin used it to further his goals, reworked its membership to limit its freedom of action, but left its framework intact. The next Soviet leader, Nikita Khrushchev, who was in power between 1953 and 1964, saw in the Academy a seed of counter power and this irritated him. He tried to change it, but he was not strong enough to do so, hence he threatened to abolish it. This is when the President of the Academy, Aleksandr Nesmeyanov, quipped, "So this is the way it is; Peter Alekseevich [the tsar] created the Academy of Sciences, and Nikita Sergeevich [Khrushchev] is going to destroy it."[1] Khrushchev soon went, and the Academy stayed. The laws in 2013 and 2014 introduced changes, the kind that Stalin did not need to consider and that Khrushchev might have only dreamt of.

The Russian Academy of Sciences was founded in 1724 in St. Petersburg. It was the St. Petersburg Academy of Sciences, then, the word Imperial figured in its name from 1747 until 1917. Some leading academicians and the Soviet leader Vladimir Lenin both recognized the mutual benefits of preserving the Academy in the new order. For a time, the Academy was under the department of mobilization of scientific forces of the ministry of education. Lenin promised political and financial support to the Academy and the academicians promised to assist in the reconstruction and development of the national economy. From 1925, the organization was renamed the Academy of Sciences of the USSR, in short, the Soviet Academy of Sciences (SAS), and it operated independently from any ministry.

The Academy of Sciences and most of its research institutes moved from Leningrad to Moscow in 1934. Some that had stayed in Leningrad, moved to Moscow as they were returning from evacuation during WWII. A whole academic city has developed in Moscow on the eastern side of Leninsky Avenue.

Physical Institute — FIAN

FIAN is the Russian abbreviation of the Physical Institute of the Academy of Sciences (*F*izichesky *I*nstitut *A*kademii *N*auk); it is named after Petr N. Lebedev. FIAN was founded in the early 1930s in Leningrad at the initiative of the world-renowned researcher in optics Sergei I. Vavilov. Many years after FIAN had moved to Moscow, it received its present home in 1951. It was built in the classical style, similar to other academic research institutes built at the same time.

---

[1] N. I. Kuznetsova, ed., *Chelovek, kotory ne umel bit ravnodyshnim: Yury Timefeevich Struchkov v nauke i zhizni* (The man who was unable to be indifferent: Yury Timofeevich Struchkov in science and in life, Moscow: Russian Academy of Sciences, 2005), p. 126.

Left: The main building of FIAN at 53 Leninsky Avenue with Petr N. Lebedev's bust in its front. *Right*: Bust of Sergei I. Vavilov in front of FIAN's fenced campus on Leninsky Avenue.

Petr N. Lebedev (1866–1912) was an experimental physicist. One of his best known discoveries was the experimental evidence he provided for James C. Maxwell's suggestion about the pressure by light on solid bodies. Lebedev graduated from the predecessor of today's Bauman University (Chapter 4) and continued his training in Germany. He joined the professorial staff of Moscow University in 1892. He developed sophisticated experiments from rudimentary components — in this, his approach was similar to that of Ernest Rutherford. Lebedev established a successful physical laboratory, which would eventually develop into the world-renowned physical institute. He was among those professors who left Moscow University in 1911 in protest against the arbitrary and illiberal policy of the minister of education, Lev A. Kasso, who violated university autonomy. Lebedev moved to the A. L. Shanyavsky City People's University (in short, Shanyavsky University) where he created another physical laboratory using his own means. Lebedev's premature death terminated his promising career.

Sergei I. Vavilov (1891–1951) graduated from Moscow University in 1914 majoring in physics. He served in WWI without stopping scientific work. Between 1918 and 1932 he taught physics at Bauman University and elsewhere. In 1926, he visited the Physical Institute of the University of

Berlin and in 1935 he toured several laboratories of optics in Western Europe. In 1933, he attended a conference on nuclear physics in Leningrad with the participation of such international physicists as Irène and Frédérick Joliot-Curie and Paul Dirac. Thus, the Leningrad physicists were early on active in nuclear physics, and in 1933 they could still have interactions with foreign scientists.

In 1932, Vavilov was appointed director of the State Optical Research Institute in Leningrad and he was also head of the physical section of the rather small physical and mathematical research institute of the Academy of Sciences. This institute expanded rapidly and was split in 1934 into the Lebedev Physical Institute and the Steklov Mathematical Institute (MIAN). Vavilov was appointed head of the Physical Institute. When he moved to Moscow, he did not entirely give up his involvement in Leningrad scientific life. For a short while he was in charge of a section of the Institute of the History of Science and Technology.

Vavilov's career continued unabated while his previously most successful biologist brother, Nikolai Vavilov (see below), was arrested and, following years of persecution, perished in prison in 1943. Sergei Vavilov tried, unsuccessfully, to help his brother. Sergei Vavilov was elected President of the Academy in 1945 and stayed in this position until his death.

In 1934, under Sergei Vavilov's research direction, his associate, Pavel A. Cherenkov, discovered what is called the Vavilov-Cherenkov Effect in the Russian scientific literature and Cherenkov Effect in the international scientific literature. For this discovery, Vavilov and Cherenkov as well as Ilya M. Frank and Igor E. Tamm, for its theoretical interpretation, received the Stalin Prize in 1946. Twelve years later, Cherenkov, Frank, and Tamm received the Nobel Prize for this discovery.[2] In addition to the recognitions during his life, Sergei Vavilov received two posthumous Stalin Prizes, one in 1951 and the other in 1952.

There are two memorial plaques on the façade of the main building of FIAN, honoring Andrei Sakharov and Igor Tamm.

When Igor E. Tamm (1895–1971) graduated from high school his parents sent him for one year to Edinburgh, Scotland. They were afraid that he would get mixed up in a political turmoil. Upon his return in 1914, he enrolled in Moscow University to major in physics and mathematics. He opposed the war but served in the Russian Red Cross. He was politically active, but not after

[2] There was yet another Soviet Nobel Prize in 1958; Boris Pasternak received it in Literature. However, the Soviet authorities blackmailed him to decline it.

Left: Memorial plaque of Igor E. Tamm: he worked here 1952–1971. *Right*: Memorial plaque of Andrei D. Sakharov: he worked here 1945–1950 and 1969–1989.

1922, because the brutality of the Bolshevik dictatorship turned him away from politics. He spent another year in Western Europe in 1927–1928, this time being sent by the State in the framework of re-establishing scientific relations. At this time international interactions were still encouraged. In the time of Stalin's terror from the mid-1930s, not only did this change; countless people fell victims to false accusations and perished. Tamm's younger brother was among them.

Tamm started his career at FIAN in 1934; organized a theoretical section there, and stayed with it for the rest of his life. He was closely associated with Leonid Mandelshtam whose main venue was the Faculty of Physics of Moscow University. Tamm joined the atomic bomb project in 1946 and worked on the hydrogen bomb in the secret installation, Arzamas-16, between 1950 and 1953.

Tamm was not only an important physicist but also a moral example for his peers. He was deeply concerned about the sad state of affairs of biology in the Soviet Union, and initiated a seminar about recent advances in biology for a group of scientists. At the beginning they held it in private rooms of academicians; just as if it were a clandestine movement, which in a way it was. In 1958, the seminar developed into a section of radiobiology in Igor V. Kurchatov's Institute of Atomic Energy. There is a Tamm Square opposite

FIAN off Leninsky Avenue (between 44 and 52 Leninsky Avenue) and one day there may be a Tamm memorial there.

Andrei D. Sakharov (1921–1989) was one of Tamm's pupils. The gap of 1950–1969 in the years on the memorial plaque at FIAN corresponds to his years at Arzamas-16. Sakharov has been labeled the father of the Soviet hydrogen bomb. He was also the Nobel Peace Prize laureate of 1975 for his activities in the protection of human rights. There was another gap between 1980 and 1986 not spelled out on the memorial plaque: Sakharov spent these years in internal forced exile under very trying conditions.

Sakharov studied at Moscow University and majored in physics. During WWII he worked parallel to his studies. After the war, he started his research career with Tamm, and in 1950 he moved with his mentor to Arzamas-16 and stayed on even after Tamm returned to Moscow. During Sakharov's two decades at the secret installation, that is, in isolation from the world of science, he tried to stay informed and continue his fundamental research. He worked on the relationship between matter and anti-matter and the evolution of the universe as well as on controlled thermonuclear reactions.

Left: Andrei D. Sakharov's statue with his bound hands behind his back (L. K. Lazarev, 2003) in St. Petersburg. *Right*: Sakharov's grave in the Vostryakovskoe Cemetery in Moscow (courtesy of Aleksandr Verny). Sakharov's second wife, the human rights activist Elena Bonner, is buried in the same grave.

Sakharov was dedicated to his work on the most effective defense of the Soviet Union. He started thinking critically about the official policy when he found further enlargements of the hydrogen bombs unnecessary and that more nuclear tests were not only superfluous but dangerous for the population. Eventually, he was eased out of the nuclear project and returned to FIAN. He was shaking the Soviet regime at its cores, but he could not be accused credibly of being anti-Soviet because the Soviet superpower status owed to him more than anybody else.

Sakharov's continuous struggle against human rights violations was such a thorn for the Soviet leadership that he was exiled to the closed city of Gorky (earlier and now again, Nizhny Novgorod). His colleagues at FIAN from time to time secured permission to visit him to maintain his connection with physics. The KGB did everything it could to make Sakharov's life unbearable in Gorky, yet he persevered. Eighteen months after Mikhail Gorbachev had come to power, Sakharov was allowed to return to Moscow. Sakharov traveled a great deal; he was elected to the Soviet parliament, and worked on his books. He died in December 1989. There is more about Sakharov's memorials in Chapter 8.

In the main building of FIAN, there are busts, reliefs, and memorial tablets honoring historic figures of science and former associates of the Institute.

Reliefs (*from the upper left corner, from left to right*): Leonhard Euler (1707–1783; the great Swiss mathematician and polymath spent much of his career in St. Petersburg), Mikhail Lomonosov, Nikolai Lobachevsky, Dmitry Mendeleev, and Petr Lebedev.

There are three memorial plaques on the wall (not shown here) for the following three physicists: Bentsion M. Vul (1903–1985) was a condensed-state physicist, a graduate of the Kiev Polytechnic Institute. He worked at FIAN from its start. He co-authored discoveries in laser physics and worked on the first laser built in the Soviet Union. Vladimir I. Veksler (1907–1966) was an experimental physicist who worked at FIAN from 1937. He graduated from the Moscow Energy Institute, investigated cosmic rays, suggested a new principle for developing accelerators, and discovered a new elementary particle. Moisei A. Markov (1908–1994), a graduate of Moscow University, was a theoretical physicist who worked at FIAN from 1934. His theoretical work focused on already known elementary particles and predicted not yet discovered ones.

Two busts in the staircase: Nikolai G. Basov (*left*) and Dmitry V. Skobeltsyn (*right*).

The staircase is decorated by two busts. Nikolai G. Basov (1922–2001) participated in WWII as a medical orderly. After demobilization, he studied at the Moscow Institute of Physical Engineering from which he graduated in 1950. As early as in 1948 he worked at FIAN under the mentorship of Mikhail A. Leontovich and Aleksandr M. Prokhorov. Basov was in leadership positions of FIAN from 1958 to 1989, first as deputy director, then, director. His field of research was quantum electronics and its applications. In 1964 he shared the Nobel Prize in Physics for his contributions to the development of masers and lasers. Prokhorov and the American Charles H. Townes were his co-laureates.

Dmitry V. Skobeltsyn (1892–1990) was a specialist in cosmic rays and the physics of high energies. He graduated from Petrograd University in 1915 majoring in physics. He worked at the Institute of Physical Technology in Leningrad between 1925 and 1939 and in the laboratory of Marie Sklodowska-Curie in Paris between 1929 and 1931. He was a professor at Moscow University from 1940. Between 1946 and 1948 he represented the Soviet Union in New York on the United Nations commission for the control of atomic energy. From 1948 until 1960, he was the founding director of the Research Institute of Nuclear Physics of Moscow University, which was named after him in 1993.

Photos of the seven Nobel laureates of FIAN, *from left to right*: Cherenkov, Tamm, Frank, Basov, Prokhorov, Sakharov, and Ginzburg.

Pavel A. Cherenkov (1904–1990) graduated from Voronezh University in 1928 majoring in physics. In 1938, Cherenkov's father was executed for alleged anti-Soviet agitation. Cherenkov completed his investigations for his scientific degrees at the Institute of Physics and Mathematics in Leningrad under Sergei Vavilov's mentorship. In the framework of their joint work, Cherenkov discovered the Vavilov-Cherenkov Effect. For the rest of Cherenkov's career he was associated with FIAN in Moscow. He was also professor at the Moscow Energy Institute and at the Moscow Institute of Physical Engineering. In 1958, Cherenkov was co-recipient of the Nobel Prize in Physics together with Tamm and Frank.

Ilya M. Frank (1908–1990) graduated from Moscow University as a physicist in 1930. His first employment was in the State Optical Institute in Leningrad. From 1934 Frank was associated with FIAN and he also became professor at Moscow University. In 1958, he shared the Nobel Prize in Physics.

In 1934, the Laboratory of Oscillations was established at FIAN and it was headed by Nikolai D. Papaleksi, Leontovich, and from 1954 until 1998, by Prokhorov. Basov was also a member of this Laboratory. In 1982, a separate Institute of General Physics was organized on the basis of this Laboratory and a couple of other laboratories, under Prokhorov's directorship. The Institute was named after him in 2002.

Left: Aleksandr M. Prokhorov's memorial plaque on the façade of the Prokhorov Institute, 38 Vavilov Street. *Right*: Prokhorov memorial (Ekaterina Kazanskaya, 2015) on Prokhorov Square — the intersection of Leninsky Avenue and Universitetsky Avenue.

Aleksandr M. Prokhorov (1916–2002) was born in Australia where his family fled from the tsarist regime. In 1923, they returned to the Soviet Union. Prokhorov graduated in physics from Leningrad University in 1939. He participated in WWII first as an infantryman, then, in reconnaissance. He was heavily wounded in 1944. From 1946 until 1982 he worked in FIAN and he remained an associate of FIAN until 1998. From 1982 he was the founding director of what is now the Prokhorov Institute. He was also professor at Moscow University and of the Moscow Institute of Physics and Technology. In 1964 Prokhorov shared the Nobel Prize in Physics.

It is a stain on Prokhorov's vita that he was one of the four academicians who in 1983 published a letter entitled "When honor and conscience are lost," condemning Andrei Sakharov in the most damning terms. It was at the time when Sakharov was in exile under constant harassment by the KGB. Publishing letters of condemnation was a favorite technique of the Soviet authorities to discredit individuals fighting for human rights. There were usually many signatories some of whom may have volunteered to sign, others may have been cajoled to do so, and there were some who did not even know what was in the letter under which their signatures appeared. In this case the very small number of signatories may have indicated that the letter was genuine. There have been recent signs of cherishing Prokhorov's memory more than some other physicists of commensurable scientific output.

Vitaly L. Ginzburg (1916–2009) gained admittance only to a correspondence course rather than the normal day course at the Faculty of Physics of Moscow University. His entrance examination was not among the best and it did not help either that he did not have a proletarian background. The next year he qualified to continue in the day course. By then he had missed the astronomy course and he never learned the elementary things about the stars in the sky. In contrast, he would make milestone discoveries about "sophisticated" astronomy, about quasars, pulsars, radio waves, X-rays, and gamma rays related to the sky.

He started in experimental optics in Leonid Mandelshtam's School at Moscow University, then, moved on to theory in Tamm's School at FIAN. Ginzburg suffered his share from the anti-Semitic campaign of Stalin's last years and was saved from the worst by proposing a crucial idea for the development of the Soviet hydrogen bomb. It was one of the three deciding ideas that made the Soviet hydrogen bomb possible. The other two ideas came from Sakharov.

Ginzburg's interaction with Lev Landau facilitated the development of one of Ginzburg's main research lines concerning superconductivity and superfluidity. His paper in 1950 in this field brought the 87-year old Ginzburg in 2003 a share of the Nobel Prize in Physics, together with Aleksei A. Abrikosov and the American Anthony J. Leggett.

Ginzburg cared for the totality of physics and its future. From time to time he enumerated the main tasks facing physics. He had an all-Moscow weekly seminar for 45 years, which he closed on the occasion of its 1700th sitting, when he was 85 years old. On the occasion of his 80th birthday he received the award "for services to the Fatherland" Class III. Ginzburg wondered whether his services or the Fatherland were labeled third class and whether he would receive Class II at 90 and Class I at 100. Actually, he did receive Class I on the occasion of his 90th birthday. Apparently the Nobel Prize helped him skip Class II.

Aleksandr L. Mints's memorial plaque with bas relief on the façade of the Institute of Radio Technique, 10 March 8 Street (courtesy of Olga Dorofeeva).

The Mints Institute of Radio Technique grew out of FIAN where initially a laboratory for radio technique was organized in 1946. This laboratory became an independent institute of the Academy of Sciences in 1957. Aleksandr L. Mints (1895–1974) was the head of the laboratory and later the director of the Institute. As Mints' activities evolved from his FIAN laboratory, its profile embraced broad areas of telecommunication, with focus on defense. He was an engineer-colonel. It was typical of Soviet science that emphasis was on the military applications rather than the transfer of technological advances for the consumer. Mints directed the building of the most powerful stations for radio transmission. He founded an institution of higher education for radio-engineering and initiated the development of a nationwide system of sensors for early warning of a possible rocket attack. Today, the Mints Institute is part of a large consortium comprising of several high tech companies.

Mints graduated from Moscow University majoring in physics. Simultaneously he was engaged in independent research at Shanyavsky University under the mentorship of Professor Petr P. Lazarev. Mints contributed creatively to the Soviet military success in the Civil War and in WWII.

Following a study trip to the United States in 1938, he was arrested on trumped up charges of anti-Soviet activities, including espionage. In 1940, he was sentenced to 10 years of slave labor, but he was freed the next year, right before the German aggression against the Soviet Union. From 1946, he participated in the nuclear project, and worked on accelerators and thermonuclear synthesis. In the 1950s, he started research on anti-ballistic missile defense and the control of Space.

Igor Kurchatov and His Institute

Monumental bust of Igor V. Kurchatov (I. M. Rukavishnikov, 1971), Kurchatov Square, in front of the Research Center "Kurchatovsky Institute." The popular reference to this memorial is "The Head."

The origin of Kurchatov's Institute can be traced back to December 1932 when Abram I. Ioffe (1880–1960) organized a group of physicists to investigate the atomic nucleus at the Leningrad Institute of Physical Technology (today, the Ioffe Institute). He appointed Igor Kurchatov to be head of the group. In 1933, the group developed into a section of the Institute and in 1944 into a division. During WWII, at the end of September 1942, the country's highest authority, the State Defense Committee, created a special laboratory of nuclear physics with Ioffe in its charge. The goal of the laboratory was to find out whether or not it was possible to develop the atomic bomb. Three physicists of the Leningrad Institute were engaged in this project: Kurchatov, Abram I. Alikhanov, and Gregory N. Flerov. Even they were not involved full time as they had been busy with other tasks as well.

The work moved to Moscow and was intensified from spring 1943; by then, Stalin had appointed Kurchatov to be the scientific head, according to Ioffe's recommendation. The infamous and highly efficient interior minister and head of the secret police, Lavrenty P. Beriya (1899–1953), was the supreme leader of the project. At this time, already ten physicists were involved in this work and other laboratories had also been involved. The secret Laboratory No. 2 of the Academy of Sciences as umbrella organization was established in April 1943. Soon, the first Soviet nuclear reactor began its operations and the secret laboratory, Arzamas-16, was established. In spring 1949, there was another name change; Laboratory No. 2 was renamed Laboratory of Measuring Instruments. Then, in 1956, it became the Institute of Atomic Energy, which in 1960 was named after Kurchatov. Today, it is the National Research Center "Kurchatovsky Institute."

Everything about Kurchatov appears to be larger than life: His memorial, "Split atom" (V.A. Avakyan, 1986) in Chelyabinsk (photo by Oleg Igoshin; courtesy of Ekaterina Bartashevich). The second Soviet nuclear weapons laboratory, "Chelyabinsk-70," was created there.

Igor V. Kurchatov (1903–1960) supervised the entire Soviet nuclear project, hence his popular label, "atomic tsar." He studied physics at the Crimean University in Simferopol.[3] Its professorial staff at one time or another

[3] After the annexation of the Crimea in 2014, it became the Vernadsky Crimean Federal University; before, it was the [Ukrainian] Vernadsky Taurida National University.

included Aleksandr Baikov, Abram Ioffe, Igor Tamm, and Vladimir Vernadsky. In 1923, Kurchatov joined Ioffe's Institute of Physical Technology. In 1940, two of Kurchatov's graduate students, Flerov and Konstantin A. Petrzhak, discovered spontaneous fission of uranium producing neutrons and much energy. When a Soviet specialist did not find anything like that in the Western scientific literature, the conclusion was that the Soviet scientists must have erred — the reviewer could not believe that a Soviet laboratory had produced an original discovery.

After the German aggression against the Soviet Union on June 22, 1941, nuclear research was halted. Kurchatov addressed himself to the more immediate task of demagnetizing ships in order to protect them from magnetic mines. When intelligence reports reached the Soviet leadership about nuclear experiments in the West for creating a new kind of bomb of heretofore unprecedented power, Soviet nuclear research was resumed.

When Stalin named the young Kurchatov to be head of the Soviet nuclear efforts in 1943, Kurchatov was not yet an academician. To counterweigh his youth, Kurchatov started growing a beard, which he vowed that he would not cut until Soviet nuclear weapons would be created. By the time those weapons had come to existence, everybody had become used to it; he kept the beard all his life, and it earned him the nickname "the Beard."

Kurchatov reported directly to Beriya and Kurchatov was the only scientist who received the complete intelligence coming from the American nuclear laboratory, Los Alamos. A legend was born about his almightiness that he always knew the solution to every problem during the project of the first Soviet atomic bombs.

The first Soviet atomic device was tested on August 29, 1949, in the Semipalatinsk proving ground in eastern Kazakhstan. On August 12, 1953, the first Soviet semi-hydrogen bomb was tested also in Semipalatinsk. The world's first nuclear power plant began its operation on June 27, 1954, in Obninsk, about one hundred kilometers (60 miles) southwest of Moscow. Kurchatov received the "Hero of Socialist Labor" award three times, one each in 1949, 1951, and 1954. He was elected academician in 1943, skipping the corresponding membership — this demonstrated Stalin's direct interference in the Academy elections. Kurchatov wielded enormous influence over matters in the Science Academy and in higher education. His influence did not diminish after Stalin's death. Kurchatov died much prematurely as a consequence of thrombosis — he had had two strokes before the fatal third.

Busts of Igor Kurchatov (*left*) and Anatoly Aleksandrov in the Kurchatov campus (courtesy of Aleksandr Verny).

Anatoly P. Aleksandrov (1903–1994) was a much decorated physicist and science administrator, and President of the Soviet Academy of Sciences between 1975 and 1986. His fields of interest were condensed-state physics, nuclear science and energetics, and polymer physics. He was directly involved with the design of nuclear power plants. Following the civil war and various odd jobs, Aleksandrov graduated from Kiev University in 1930. He participated in scientific research in Ioffe's Institute. When WWII began for the Soviet Union, Aleksandrov worked on the protection of ships from magnetic mines, which brought him together with Kurchatov. Aleksandrov started working on the atomic bomb project in 1943. When Petr Kapitsa was fired from his Institute of Physical Problems (see below), Aleksandrov was appointed in his stead from 1946 until 1955. In 1955 Aleksandrov became Kurchatov's deputy and when Kurchatov died, Aleksandrov was appointed director of the Kurchatov Institute. At the time of the Chernobyl catastrophe, in 1986, Aleksandrov was both the President of the Soviet Academy of Sciences and the Director of the Kurchatov Institute of Atomic Energy; thus he carried double responsibility. On November 30, 1975, there was an accident at the Leningrad Nuclear Power Plant, which could be considered Chernobyl's forerunner. Thus, the lesson was offered, alas, it was not learned.

The Leningrad and the Chernobyl accidents sadly enveloped Aleksandrov's presidency of the Academy of Sciences. After Chernobyl, Aleksandrov fell out of favor with the Soviet leadership; he resigned from the presidency of the Academy in 1986 and from the directorship of the Kurchatov Institute in 1988. In the Soviet regime, old age would not have been reason for such resignation. Whereas before 1986, he was showered with the highest awards, he was given only a modest award on the occasion of his 90th birthday. When he died, he was accorded a state funeral, but was buried in the Mitinskoe cemetery (Chapter 8) rather than in the Novodeviche as some sources state.

Bust of Lev A. Artsimovich (courtesy of Aleksandr Verny).

Lev A. Artsimovich (1909–1973) graduated from Belarusky University in Minsk in 1929. From 1930, he worked in Ioffe's Institute. When the Institute initiated research in nuclear physics, Artsimovich was among the first to join it. When these studies moved to Moscow, he went with them. His main contributions to the project included methodologies for isotope separation and studies in controlled thermonuclear reactions.

In 1966, Artsimovich signed the so-called "Letter of 25," which was a letter by scientists, writers, and artists asking the then supreme leader Leonid Brezhnev not to rehabilitate Stalin. There was growing concern that the danger of Stalin's rehabilitation was real. Among the 25 signatories, there were five scientists, viz., Artsimovich, Kapitsa, Leontovich, Sakharov, and Tamm. Brezhnev left the letter unanswered, but the next congress of the Communist Party did not discuss any change in the decisions of previous congresses condemning Stalin's crimes. Artsimovich recognized the almightiness of the Soviet establishment in all matters of science. He noted, "Science exists in the palm of the State and the warmth of this palm keeps science alive."[4]

A number of additional physicists and technologists have memorial plaques in the Kurchatov campus (not shown). Mikhail A. Leontovich (1903–1981) started his college education at Shanyavsky University and completed it at Moscow University in 1923 majoring in physics. First he worked in the Institute of Biophysics, then joined Mandelshtam at the University where Leontovich's research focused on molecular optics. He continued his research at FIAN and became professor of physics at Moscow University and at the Moscow Institute of Physics and Technology. He distinguished himself with his theoretical studies in radiophysics. From 1951 Leontovich was in charge of the theoretical work on thermonuclear synthesis at Kurchatov's Institute.

Leontovich was already a corresponding member of the Science Academy when he was proposed for full membership at the first post-war elections to the Academy in 1946. Igor Tamm was also due for full membership, but Tamm's candidacy was eliminated by the representative of the Communist Party. At this point Leontovich let the President of the Academy know that he, that is, Leontovich, did not want to become full member if it would mean taking Tamm's deserved place. Leontovich could not have it his way, but he showed an example of selflessness.

Boris B. Kadomtsev (1928–1998), a physicist, graduated from Moscow University in 1951 and joined Kurchatov's Institute in 1956. He worked on theoretical problems of the nuclear reactors under Dmitry I. Blokhintsev. Later, Kadomtsev's research concerned plasma physics and controlled thermonuclear synthesis. He was professor at the Moscow Institute of Physics and Technology and edited the influential periodical *Uspekhi fizicheskikh nauk* (*Advances in Physics*) between 1976 and 1998.

Mikhail D. Millionshchikov (1913–1973) graduated as oil engineer from the Grozny Petroleum Institute in 1932 (Grozny today is the capital of the Chechen Republic). Following a variety of jobs, he joined Kurchatov's Institute and stayed with it for the rest of his career. He was professor of molecular physics at the Moscow Institute of Physical Engineering and professor of aeromechanics at the Moscow Institute of Physics and Technology. He had high-level appointments in state and scientific governing bodies and organizations.

---

[4] Mark Popovsky, *Upravlyaemaya nauka* (Regulated Science, Free electronic library, royallib.ru [accessed July 23, 2017])

Yuly B. Khariton worked for the Kurchatov Institute from 1944 until 1949. There is more about him below.

Valery A. Legasov's statue (erected in 2016) at his former school and his bust (Nikolai and Vasily Selivanov, 2017) at the Faculty of Chemistry, Moscow University. Courtesy of Olga Dorofeeva.

Valery A. Legasov (1936–1988), an inorganic chemist, worked in the Kurchatov Institute from 1962 until 1988. He had a brilliant career, but many brilliant scientists worked in the Kurchatov Institute. The very last period of Legasov's life and his demise added an extraordinary dimension to his career.

He attended School No. 56 at 22 Kutuzovsky Avenue. This school now bears his name and his statue stands on the school grounds. Legasov graduated from the Mendeleev University of Chemical Technology and was a professor there from 1978. He had another professorial appointment at Moscow University from 1983, and there is now his bust next to the lecture hall named after him. He was young when he became an academician and from 1983, he was first deputy director of the Kurchatov Institute of Atomic Energy. He had a broad research program involving nuclear techniques and another in the area of noble gases. Yet another line of his interest was, most significantly, in nuclear safety for which he advocated new criteria, new technologies, and new guidelines.

When the catastrophe at the Chernobyl Nuclear Power Plant happened in 1986, Legasov was appointed to the government commission charged with mitigating the after-effects of the accident. He was among the first who appeared on the scene and spent four months there, much more than he was supposed to. Although it was already at the time of Mikhail Gorbachev's "glasnost — openness," the Soviet Union was not at all forthcoming with information about what happened and what was going on. In contrast, Legasov, even at an international forum, discussed openly issues that his superiors considered classified. There were repercussions and Gorbachev declined awarding him the Hero of Socialist Labor title, which Legasov had richly deserved. On the second anniversary of the Chernobyl catastrophe, Legasov hanged himself. In his suicide recording, he revealed heretofore unknown information about the accident, which was then incorporated into a BBC film about Chernobyl. In 1996, Legasov was posthumously awarded the Hero of Russia title for his courage, tenacity, and heroism.

Bust of Isaak K. Kikoin in Building 103 (courtesy of Aleksandr Verny).

Isaak K. Kikoin (1908–1984) studied at the Leningrad Polytechnic Institute and while being a student he started doing research at Ioffe's Institute. Upon graduation in 1930, Ioffe arranged for Kikoin a study trip to Western Europe. Back in Leningrad, he continued research with Ioffe and became a member of the professorial staff.

In 1943, he was among the first physicists who joined the new Laboratory No 2 charged with the development of the atomic bomb. In spring 1945, he was in the group of scientists sent to Germany to locate German scientists who could be useful for Soviet programs and find instrumentation and raw materials for the nuclear project. Yuly Khariton and Lev Artsimovich were also among the members of this special group. They interviewed German scientists and succeeded in collecting about one hundred tons of uranium ore. This accelerated the Soviet nuclear program. In the 1950s, Kikoin was in charge of the development of methodology for remote registration of nuclear explosions. He remained active in research and in the leadership of Kurchatov's Institute.

A few more scientists are mentioned here that are honored with memorials in the Kurchatov campus. Boris V. Kurchatov (1905–1972) was Igor's brother. He graduated as a research chemist in 1927 from Kazan University. He worked at Ioffe's Institute between 1928 and 1943. Then, he joined Laboratory No 2 and stayed with the nuclear project for the rest of his life. His research was in radio-chemistry and in condensed-state chemistry and physics.

Vladimir D. Rusanov (1929–2007) graduated as a physicist from Moscow University. He worked at FIAN, then moved to Kurchatov's Institute. He was a world authority of hydrogen energetics. His research program developed into the Institute of Hydrogen Energetics and Plasma Technologies within the framework of the Research Center "Kurchatovsky Institute." He was head of department of plasma physics and chemistry at the Moscow Institute of Physics and Technology.

Evgeny K. Zavoisky (1907–1976) graduated as experimental physicist from Kazan University in 1931. He continued his studies and in time became the head of the department of experimental physics at his Alma Mater. Between 1947 and 1951 he worked at Arzamas-16. From 1951, he was an associate of Kurchatov's Institute.

Zavoisky is the world-renowned discoverer of electron paramagnetic resonance, a physical technique for the investigation of the structure of materials. For this achievement he could have received the Nobel Prize. One of the reasons he did not may have been his isolation from the international community, which was generally characteristic of Soviet science. In his case it was compounded by his working in a closed institution. According to the late Nobel laureate Vitaly Ginzburg, Zavoisky deserved the Nobel Prize more than anybody among those Soviet scientists who might have won it, but never did.

Georgy A. Gladkov (1925–2005) was a graduate of the Moscow Energy Institute. After his post-graduate studies he worked on the development of industrial reactors producing plutonium at Kurchatov's Institute. He participated in the project of developing nuclear weapons and reactors, in particular those deployed for the Soviet Navy, including nuclear submarines and icebreakers. His specialty was energetics; he researched thermal physics and energy transportation. He was professor at the Moscow Institute of Physical Engineering.

Vladimir I. Merkin (1914–1997) graduated from the Moscow Institute of Chemical Machine Construction. From 1944 he worked in Laboratory No. 2 developing constructions of the gun method for the atomic bombs. In this, one subcritical mass of fissionable material is shot into another subcritical mass of fissionable material, thus producing a supercritical mass, and hence, an explosion. He worked for the Navy and other areas of the armed forces, the ship industry, and spaceships, as well as for nuclear reactors for energy production. He stayed with Kurchatov's Institute to the end of his life.

There is a memorial tablet devoted to 11 experts of nuclear science and technology. They participated in creating the first nuclear reactor on the Eurasian continent, which was put into operation on December 25, 1946. All of them had received State and Lenin Prizes (the State Prize was originally the Stalin Prize): S. A. Baranov, V. V. Goncharov, V. A. Davidenko, V. S. Obukhov, I. S. Panasyuk, N. F. Pradyuk, Yu. A. Prokofiev, D. L. Simonenko, G. A. Stolyarov, and G. N. Flerov. The world's first nuclear reactor was put into operation in Chicago on December 2, 1942. The Americans were unaware of the Soviet advances and this is why the news about the first atomic explosion in August 1949 was met with so much disbelief in the United States.

Mathematical Institute — MIAN

Left: The Steklov Institute of Mathematics of the Academy of Sciences (MIAN), 8 Gubkin Street. *Right*: Ivan M. Vinogradov's memorial in the Institute.

Already Leonard Euler formed a mathematical cabinet in 1766 in St. Petersburg in the framework of the Academy of Sciences. After WWI and the revolutions, almost everything had to be started all over. In 1919 Vladimir A. Steklov (1864–1926) initiated a new Mathematical Cabinet of the Academy of Sciences. In 1921, he was one of the initiators of the Physical-Mathematical Institute. Upon Steklov's death, the Institute was named after him. The mathematical section of the Institute grew into the Steklov Mathematical Institute (MIAN) and the physical section into the Lebedev Physical Institute (FIAN).

Ivan M. Vinogradov (1891–1983) was the director of MIAN for half a century. FIAN and MIAN came to existence in about 1934 and almost immediately they both moved to Moscow. Vinogradov's directorship acquired international notoriety for anti-Semitism that manifested itself in his discriminatory hiring practices. The Soviet authorities practiced ill-masked anti-Semitism, especially in hiring, but Vinogradov went even further than what was expected of him.

Mstyslav V. Keldysh memorial in front of the Keldysh Institute, 4 Miusskaya Square.

In 1966 a new institute was spun off from MIAN; it is now the Keldysh Institute of Applied Mathematics (there is more about Keldysh in Chapter 4).

The origins of the Keldysh Institute reach back to 1953 when a special division of MIAN was established to assist ongoing government programs. They included the nuclear project, the conquest of space, missile defense, and fusion research. More recently computational biology and robotics have been added to the profile of the Keldysh Institute.

Anatoly A. Dorodnitsyn's memorial tablet on the façade of the Dorodnitsyn Computing Center, 40 Vavilov Street.

The Dorodnitsyn Computing Center grew out of applied mathematics in 1955. Its founding director, Anatoly A. Dorodnitsyn (1910–1994), worked in this institution between 1955 and 1994.

The Siberian Branch of the Academy of Sciences was also a spinoff from the Steklov Institute. Its two principal initiators in 1957 were Mikhail Lavrentiev and Sergei Khristianovich.

Mikhail A. Lavrentiev (1900–1980) was a mathematician with affiliations both at the Steklov Institute and at Moscow University. He graduated from Moscow University and stayed on for postgraduate studies under the internationally renowned Nikolai Luzin's mentorship. Afterwards, Lavrentiev was sent for a study trip to France. He taught at various institutions, among them the Bauman University and the Zhukovsky Institute of Aero-hydrodynamics and worked at the Institute of Mathematics of the Ukrainian Academy of Sciences. During WWII, Lavrentiev was engaged in defense-related research.

He was active in creating the Moscow Institute of Physics and Technology and directed the Institute of Exact Mechanics and Computational Technology. Between 1953 and 1955, he was deputy to the scientific director of Arzamas-16. When he arrived there in 1953, he was considered to be the replacement of Yuly Khariton, the scientific director, in case the next test would fail and Khariton would have to face the consequences of the failure. Lavrentiev was the first president of the Siberian Branch of the Academy of Sciences in Novosibirsk. His statue stands in Novosibirsk.

Sergei A. Khristianovich (1908–2000) was a mathematician with a principal interest in mechanics. He graduated from Leningrad University in 1930. He was at MIAN from 1935, then in 1939 he moved to the new Institute of Mechanics and in 1940, he moved to the Zhukovsky Institute of Aero-hydrodynamics.

Sergei A. Khristanovich's memorial plaque (K. Sinyavin) on the façade of 43 Profsoyuznaja Street where he lived between 1985 and 2000 (courtesy of Aleksei Semenov).

During the war, Khristianovich worked on defense-related problems and his suggestions enhanced the power of the legendary "Katyusha" multiple rocket launchers. He was one of the initiators and one of the first professors of the Moscow Institute of Physics and Technology. He participated in the further development of nuclear weapons in the period 1953–1961. He was concerned about the destruction a nuclear attack might bring to the highly concentrated scientific centers in the European regions of the Soviet Union. Hence, he co-sponsored the proposal of developing a scientific center in Siberia and he moved there. He returned to Moscow in 1965 and continued his research activities and organizational work. He had broad visions and the ability to convince people high in the hierarchy to make his visions become reality. This was a rare accomplishment as there was little room for private initiatives in Soviet society.

Part of the photo gallery at the Steklov Institute of members of the Academy connected with MIAN. Some of the mathematicians who figure in our book are identified by column number (x) from the left to right and by row number (y) from top to bottom, i.e., by x/y: Arnold 1/3, Bogolubov 2/3, Vinogradov 4/1, Dorodnitsyn 6/1, Zeldovich 6/2, Kantorovich 7/1, Keldysh 7/2, A. N. Krylov 7/3, Kolmogorov 8/2, Kochin 9/1, Lavrentiev 9/2, and Luzin 10/3.

The collection of photographs on the walls in the lobby on the 9th floor of the Steklov Institute indicates the enormous influence of this Institute on Soviet/Russian science, education, space exploration, and especially, defense.

Petr Kapitsa and His Institute

Main entrance to the Kapitsa Institute, 2 Kosygin Street, and the former residence of Kapitsa and his family on the site of the Institute; now it is the Kapitsa Memorial Museum.

Kapitsa's memorial plaque at the entrance to the main building.

Petr L. Kapitsa (1894–1984) studied at the St. Petersburg/Petrograd[5] Polytechnic Institute and Abram Ioffe was his mentor. Upon graduation, Kapitsa joined Ioffe in the new Roentgen Institute and soon in the new Institute of Physical Technology. Kapitsa did summer studies under Leonid Mandelshtam's mentorship. In 1921, Kapitsa traveled to England as a member of a Soviet delegation and stayed in Cambridge where he became Ernest Rutherford's disciple. Kapitsa built up a strong group and research direction to investigate magnetism at the Cavendish Laboratory.

Kapitsa returned for regular visits to the Soviet Union until 1934 when he was not let out of the Soviet Union at the end of such a visit. After a brief period of depression, Kapitsa soon came to grips with his new situation, accepted the offer of the Soviet leadership, and built up his own research institute on a choice site in Moscow. It was the Institute of Physical Problems of the Academy of Sciences. The name of the Institute indicated that there was no specific field of physics for its activities; rather, its staff would determine the fields of its inquiry. Soon Kapitsa discovered the superfluidity of helium, which Lev Landau interpreted theoretically. Kapitsa received a Nobel Prize in 1978 for this discovery and Landau in 1962 for his (see below).

---

[5]The German-sounding name St. Petersburg was changed to Petrograd in 1914, upon the outbreak of WWI, by the tsarist regime and to Leningrad in 1924 by the Soviets. It was changed back to St. Petersburg in 1991. However, the surrounding area retains the name Leningradskaya Region.

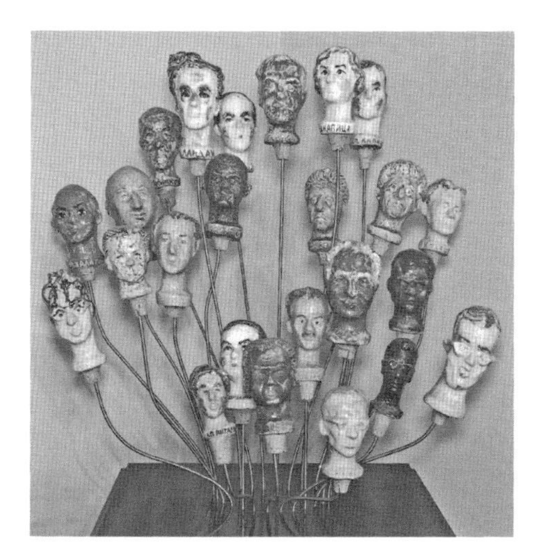

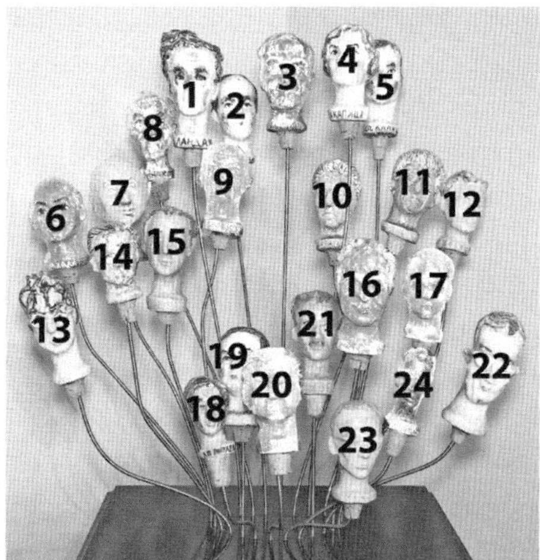

Petr L. Kapitsa and his associates — all renowned physicists in their own right — in 1964 represented by majolica stoppers. It was a present on the occasion of Kapitsa's 70th birthday. 1 Landau L. D., 2 Kapitsa A. P., 3 Kapitsa P. L., 4 Kapitsa A. A., 5 Kapitsa S. P., 6 Khalatnikov I. M., 7 Lifshits E. M., 8 Minakov N. N., 9 Alekseevsky E. E., 10 Arefiev V. V., 11 Shalnikov A. I., 12 Pavlov A. I., 13 Rubinin P. E., 14 Okolesnov S. P., 15 Borovik-Romanov A. S., 16 Malkov M. P., 17 Filimonov S. I., 18 Pitaevsky L. P., 19 Abrikosov A. A., 20 Perevozchikov V. I., 21 Khaikin M. S., 22 Vainshtein L. A., 23 Yakovlev S. A., 24 Gorkov A. P.

Kapitsa was a brave man who wrote letters to Stalin. However, even in his most critical letters concerning various aspects of Soviet life, Kapitsa appeared identifying himself with the goals of the Soviet regime. A considerable portion of Kapitsa's letters to Stalin, and also to some of Stalin's lieutenants, were for asking that important scientists be freed from incarceration, and thus, saving their lives. In 1938, Lev Landau, the most brilliant theoretician of Landau's generation, was incarcerated and was facing a bleak future — really, no future. Kapitsa took action and persuaded the authorities to let him assume personal responsibility for Landau.

During WWII Kapitsa helped the war effort by organizing oxygen production, and he relied on his own invention of a new technology. After the war, Kapitsa got into conflict with Beriya who was in charge of the atomic bomb project and Kapitsa was a member in a high ranking commission of the project. He complained to Stalin comparing Beriya to a conductor of an orchestra who had the conductor's baton in his hand and waved it, but did not have the score. Kapitsa could not know that Beriya did have the score; it was the detailed intelligence reports about the American bomb design. Kapitsa wanted to have his original ideas tested for the Soviet bomb, but there was neither time nor desire to deviate from the stolen blueprint. Even if the first Soviet bomb was a copy of the American, its production required high technical skill and rigorous adherence to the plan. All this was happening in a country devastated by the war and whose infrastructure even without the destruction had not been at a high level.

Stalin removed Kapitsa from the bomb project and even from his own Institute whose members, including Landau, were badly needed for the atomic bomb program. During his exile between 1946 and 1954, Kapitsa built up a research project in his summer home (*dacha*) under rudimentary conditions. In Kapitsa's absence, Anatoly Aleksandrov was the director of the Institute of Physical Problems. When Sergei Vavilov, the physicist president of SAS died, the Institute was named after him. Following Stalin's death and Beriya's disappearance from the scene, Kapitsa returned to his Institute. After Kapitsa's death, Vavilov's name was transferred to another institute and the Institute of Physical Problems was named after Kapitsa.

Busts of Petr Kapitsa and Anna Kapitsa (née Krylova) by N. B. Nikogosyan (1955) in the Kapitsa Memorial Museum.

Kapitsa's first wife and their newborn baby daughter died of the Spanish flu and their son had died of scarlet fever. Anna A. Kapitsa was Kapitsa's second wife; they married in 1927 and their happy marriage lasted until Kapitsa's death in 1984.

Bust of Aleksei N. Krylov in the park adjacent to the Northern River Port of Moscow at 51 Leningradsky Highway (courtesy of Olga Dorofeeva).

Anna was the daughter of academician Aleksei N. Krylov (1863–1945), a high-ranking Russian naval engineer, whose specialty was the mathematics and mechanics of shipbuilding. He was a member of the Soviet academic group sent by the Soviet government in 1921 to Western Europe to re-establish scientific ties with the West. Kapitsa was also in this group, hence their acquaintance. Kapitsa and his second wife had two sons, the physicist and popular science TV personality Sergei (1928–2012) and the geographer and explorer Andrei (1931–2011).

Memorial tablet of Lev D. Landau on the façade of the Institute of Physical Problems. According to the tablet, Landau lived and worked here from 1937 to 1968. The tablet does not mention the year of Landau's incarceration at the Lubyanka, 1938–1939.

There are hardly any memorials in Moscow honoring Kapitsa or Landau. Lev D. Landau (1908–1968) graduated from Leningrad University at the age of nineteen. Paul Ehrenfest, Aleksandr Friedman, and Abram Ioffe were among his professors. He did his post-graduate work at Ioffe's Institute. Landau first worked at the Ukrainian Institute of Physics and Technology in Kharkov, then, at Kapitsa's Institute. In the early 1930s, he did research in Western Europe, especially in Copenhagen with Niels Bohr, whom Landau considered his mentor.

Landau was arrested in 1938 and was kept locked up for an entire year with continuous interrogations. During his incarceration he maintained his sanity by working on physical problems. By the time his ordeal ended he had stored four papers in his head. Upon his release under Kapitsa's supervision (a stipulation of which Landau remained unaware to the rest of his life), he never talked about his experience. After the war, the authorities involved Landau in the nuclear project, which he quit immediately upon Stalin's death.

Landau built up a brilliant School of theoretical physics. He administered a multi-part examination of mathematics and theoretical physics, the so-called *teorminimum*. Anybody was welcome to take this exam nationwide. From his Kharkov days through 1961 a total of 43 scientists had passed the complete system of *teorminimum*. Landau's pedagogical activities included his multi-volume treatise *Course of Theoretical Physics* coauthored with Evgeny M. Lifshits (1915–1985). Lifshits was also a theoretical physicist and after Landau's death he was elected academician. It was an unfair characterization of the situation that there was not a line written by Landau and not a thought by Lifshits in the *Course*. Landau's phobia of writing and Lifshits's devotion to Landau may have fed this unfair characterization; in reality theirs was a superb partnership of creativity.

Landau received the Nobel Prize in Physics in 1962 and it was handed to him in hospital. On January 7, 1962, he suffered an almost fatal automobile accident. It was a miracle that he survived — thanks to extraordinary medical attention. Sadly, he spent his last six years under excruciating physical pain and never regained his legendary scientific prowess. In 1964, Landau's theoretical group split from the Institute of Physical Problems and a new Institute of Theoretical Physics was established in Chernogolovka, Moscow Region. It has been named after Lev D. Landau.

Nikolai Semenov and His Institute

Archival photograph of a Semenov memorial tablet, which was heavily damaged in a fire in 2012 (courtesy of Aleksei Semenov).

The Semenov Institute of Chemical Physics was established in 1931 in Leningrad on the basis of Ioffe's Institute of Physical Technology. Semenov was one of Ioffe's closest disciples. The purpose of the new institute was to apply physical theory and the methodology of physics to chemistry and to the chemical industry. The Institute moved to Moscow after the evacuation, towards the end of WWII. Its location at 4 Kosygin Street is one of the choicest sites in the capital city. It is on Vorobyovy Gory (Sparrow Hills), which was called Lenin Hills for decades under the Soviet rule. From this location, there is a spectacular view to the Moscow River and to downtown Moscow. The Institute has a big lot even though one section had already been cut from it when the Emanuel Institute of Biochemical Physics was established. Previously, Nikolai M. Emanuel (1915–1984) was head of a section in Semenov's Institute focusing on physical chemical research of biological problems.

The supreme location of the Semenov Institute may carry some hazards in today's uncertain world. As the independence of the Academy of Sciences has been cut, it is no longer the owner of the real estate where its institutes stand, not even of the institutes themselves. Thus the institutes may be at the mercy of the powers that be. What if some wealthy institution or organization, or even a wealthy individual, makes an offer to the owners of the sites that is impossible to resist? The dilapidated state of the main building of the Semenov Institute illustrates the precariousness of the economic situation that it is facing. This main building was once upon a time the summer home of the aristocrat Mamonov family.

Nikolai N. Semenov (1896–1986) wielded enormous authority in the Soviet scientific hierarchy. On his part, he revered the authority of the Soviet leadership and even the authority of the bureaucrats of the Academy of Sciences. He began his university studies at St. Petersburg University where Ioffe became his mentor. Ioffe studied in St. Petersburg and then he was W. C. Roentgen's doctoral student in Munich. Ioffe built up a large school and planted the seeds of research institutes all over the Soviet Union. Semenov emulated Ioffe, but while Ioffe sent his former disciples on their independent ways, Semenov kept everybody in his amicable fold. He built up an empire of research venues. When the American Nobel laureate chemist Robert B. Woodward invited the 80-year-old Semenov to visit his ranch of hundreds of cows, Semenov reciprocated. He invited Woodward to visit his institutes in Moscow where five thousand researchers worked under his command.

Semenov had outstanding associates. Viktor N. Kondratiev (1902–1979) was a physical chemist whose principal interest was in the mechanism of chemical reactions and the structure of matter. Upon graduation from the Polytechnic Institute, he started working in Semenov's laboratory and stayed on as one of the leading associates of Semenov's Institute for the rest of his life.

Aleksandr I. Shalnikov (1905–1986) graduated as engineer physicist from the Polytechnic Institute in 1928. Already from 1923, he worked in Ioffe's Institute of Physical Technology. He was an exceptional experimentalist and designed the most refined instruments. From 1935, he was an associate in Kapitsa's Institute and worked in low-temperature physics. Already from 1946, Shalnikov was involved in the atomic bomb project. For the 1947–1955 period — this was the time of Kapitsa's forced absence from his own Institute — it is only known that Shalnikov worked on projects of national significance and that he received three Stalin Prizes. Shalnikov returned to his previous research from 1955 and was professor of physics at Moscow University. Among many other interests, he assisted physicians in developing surgeries with frozen tissues.

Left: Memorial plaque of Yakov B. Zeldovich on the façade of the Keldysh Institute of Applied Mathematics (courtesy of Lyubov Mikhailova). *Right*: Memorial plaque of Yuly B. Khariton at the Semenov Institute (courtesy of Aleksei Semenov).

Semenov had an excellent eye to find gifted young scientists and he let them develop their own research. One of them was Yuly B. Khariton (1904–1996) who designed an experiment and made surprising observations. They served then as basis for Semenov's milestone discovery of the theory of branched chemical chain reactions. This discovery brought Semenov his share of the Nobel Prize in Chemistry in 1956. Having already proved himself an excellent experimentalist, Khariton embarked on his trip to Cambridge where he earned his PhD degree under James Chadwick's mentorship. In a few years' time Chadwick discovered the neutron in 1932.

Khariton returned to Leningrad and he and Yakov B. Zheldovich (1914–1987) jumped into researching nuclear fission when they had heard about its discovery in Germany. The principle of branched chemical chain reactions and the principle of nuclear chain reactions were analogous. After the war, Khariton became the scientific head of Arzamas-16, and Zeldovich one of its chief theoreticians. Both Khariton and Zeldovich were Jewish and their appointment to Arzamas-16 happened at the initial stage of Stalin's fierce anti-Semitic campaign soon following WWII. Arzamas-16 was a highly classified project, but Khariton and Zeldovich were so outstanding that in their cases for the sake of the success of the atomic bomb project the anti-Semitic barrier was removed. Zeldovich returned to Moscow and to pure science when he was fifty years old. He became an associate of the Institute of Applied Mathematics (see above) for the next 18 years.

Semenov's Institute was one of the centers of the atomic bomb program, and it was to his credit that fundamental research was not neglected even in the years of such accelerated efforts focusing on a few selected goals. He recognized the damage Lysenko and Lysenkoism caused to Soviet science over the years. He even had his own "Lysenko affair" in that a physics professor, Nikolai Akulov, at Moscow University, accused Semenov of plagiarizing the theory of two Danish scientists. This happened on the eve of the war, in April 1941. When Akulov was told that he was advocating the preeminence of Western scientists over a Soviet one, he changed his tune. He now accused Semenov of plagiarizing the work of a Russian physical chemist. Because of the war, this affair dragged on for years before Semenov could put it behind him, and afterwards Akulov had to move from Moscow to Minsk.

Semenov appeared sensitive to the plight of Lysenko's victims. The geneticist Iosif Rapoport was one of them and when he had no place to work, Semenov offered him a laboratory. Iosif A. Rapoport (1912–1990) graduated from Leningrad University majoring in biology and specializing in genetics. He did postgraduate studies under Nikolai Koltsov at the Institute of Experimental Biology. Rapoport defended his dissertation in the Institute of Genetics. In a few years' time he completed his dissertation for the DSc degree, but upon the German aggression he volunteered for the Red Army. His defense had to be postponed and it took place in 1943 during Rapoport's recuperation after one of his serious wounds. For his heroism in actual battles he had been nominated three times for the title Hero of the Soviet Union and each time he received a lower award.

After the war, Rapoport continued his research in genetics at the Institute of Cytology, Histology, and Embryology. He discovered chemicals that had a strong impact on mutations (chemical mutagenesis). He used *Drosophila melanogaster*, the favorite fruit fly in genetic research, in his experimental studies. These were the kinds of experiments that ignorant Soviet leaders and science administrators liked ridiculing. Rapoport bravely contested Lysenko's anti-science views at the infamous session of the Academy of Agricultural Sciences in August 1948. As a punishment, the Communist Party, which Rapoport joined during the war, excluded him from its membership in 1949. The more severe punishment was that he could no longer do research according to his interest and qualifications. He found employment in expeditions and did paleontological investigations between 1949 and 1957. It was then that Semenov offered Rapoport assistance and Rapoport returned to his research on chemical mutagenesis.

In 1962, there was a notion of nominating Rapoport for the Nobel Prize together with the German refugee British scientist Charlotte Auerbach, FRS. Rapoport's and Auerbach's names, however, did not show up among the nominations. It has been alleged that in an unprecedented move the Nobel Committee had asked the Soviet authorities about their stand if such an award would be made. The rules of the Nobel Prize exclude any interference on anybody else's part — let alone governments and political parties — in the matters of nominations and everything else. Nonetheless, the story appears plausible because of the unpleasant reception of Boris Pasternak's Nobel Prize in Literature in 1958 by the Soviet authorities. Unfortunately, cases have been known when the Nobel Foundation and the Nobel Committees tolerated, possibly welcomed, the interference of Soviet organs in the matters of nominations of Soviet scientists. Furthermore, it is known how timidly the Swedish authorities behaved vis-à-vis the Soviet Union when they should have pursued what had happened to Raoul Wallenberg.[6] Again, allegedly, the Soviets told Rapoport that he should apply for the reinstatement of his party membership upon which they would not object to his Nobel Prize. This Rapoport refused and there was no Nobel for Rapoport and none for Auerbach either.[7]

From the mid-1960s, with Khrushchev's fall, Rapoport's possibilities expanded and he started receiving official recognition. Finally, on October 16, 1990, he was awarded the title of Hero of Socialist Labor. On December 25, 1990, Rapoport was run over by a truck as he was crossing a street. He died a few days later.

There are memorial tablets honoring N. S. Enikolopov and V. I. Goldansky at the Semenov Institute (not shown here). Nikolai S. Enikolopov (1924–1993) was a Soviet-Armenian scientist, a member of Semenov's School, specializing in the research of the kinetics of chain reactions — originally one of Semenov's main interests. He founded and directed an independent research institute of synthetic polymers. Vitaly I. Goldansky (1923–2001) was an internationally renowned scientist during the last period of the Soviet era and an outspoken reformer. He was Nikolai Semenov's son-in-law and became the director of the Semenov Institute after Semenov's death.

---

[6] Raoul Wallenberg was a Swedish diplomat who saved thousands of Jews in Budapest in 1944 during the Hungarian Holocaust. During the liberation of Budapest, the Soviets arrested him in January 1945, and he was never heard of again.

[7] A.M. Blokh, *Sovietskii Soyuz v interiere Nobelevskikh premii* (Moscow: Fizmatlit, 2005); pp. 512–513. The book is available in English translation: Abram M. Blokh, *Soviet Union in the Context of the Nobel Prize* (Singapore: World Scientific, 2018).

Memorial plaque of Victor L. Talrose at the Semenov Institute (courtesy of Aleksei Semenov).

Victor L. Talrose (1922–2004) interrupted his studies at Moscow University when, following the German aggression in 1941, he volunteered for the front as a sophomore. He was wounded three times, but survived — only two percent of his age group of those who fought survived the war. He graduated in 1947 and worked for forty years in Semenov's Institute. Between 1949 and 1956, he participated in the nuclear weapons program. He was a leading professor of chemical physics at the Moscow Institute of Physics and Technology. In 1987, he organized his Institute of Energy Problems in Chemical Physics on the basis of the section he headed at the Semenov Institute. In 1997 he moved to California where he continued his research but maintained his Moscow laboratory as well. He was a pioneer in chemical lasers. He established his fame in 1952, when he and his student, Anna K. Lyubimova, observed, in today's terminology, the carbocation CH_5^+.

Memorial of Nikolai N. Semenov and Fedor I. Dubovitsky (Artem Rodionov, 2016) in Chernogolovka (courtesy of Artem Rodionov and Aleksei Semenov).

By the mid-1950s, the expansion of Semenov's Institute could no longer be limited to the Moscow site. Some of the research was about explosives for which a location outside of Moscow was desirable. Semenov chose a site about 55 kilometers (34 miles) northeast of Moscow. Subsequently, a sizzling venue of science grew at Chernogolovka where there used to be a village. The construction was directed by the physical chemist and Semenov disciple Fedor I. Dubovitsky (1907–1999).

Shubnikov Institute of Crystallography

Aleksei V. Shubnikov's bust at the Faculty of Physics of Moscow University (courtesy of Larissa Zasourskaya).

Aleksei V. Shubnikov (1887–1970) was interested in crystals already in his childhood and attended lectures at the Polytechnical Museum in Moscow, and he became a crystallographer. He graduated from Moscow University in 1912 and started his career at Shanyavsky University. He worked at various places before his appointment to be in charge of the Laboratory of Crystallography in 1937 from which the Institute of Crystallography of the Academy of Sciences (IKAN) developed. Shubnikov stayed on as director from 1944 until 1962. He has a memorial tablet on the façade of IKAN at 57 Leninsky Avenue. He was also head of the department of crystallography at Moscow University. He wrote a beautiful book on symmetry that has become a classic. Following Shubnikov, between 1962 and 1996, Boris K. Vainshtein (1921–1996) took over the directorship of IKAN. It was at the time when the Institute was moving into its new building on Leninsky Avenue. Vainshtein worked in the Institute from 1945 until the end of his life. The international interactions of the Institute, and of the entirety of Soviet/Russian crystallography greatly expanded under Vainshtein's leadership. In 1984, the UK Prime Minister Margaret Thatcher visited IKAN during her state visit in Moscow. Ms. Thatcher had been a student of the Nobel laureate British crystallographer Dorothy Hodgkin.

Kurnakov Institute

Left: Nikolai S. Kurnakov's bust in front of the Kurnakov Institute of General and Inorganic Chemistry, 31 Leninsky Avenue. *Right*: Postage stamp of 1968 issued for the 50th anniversary of the Kurnakov Institute.

The Kurnakov Institute of General and Inorganic Chemistry traces its history to Lomonosov's first scientific chemical laboratory and to the more immediate date of 1918. The Institute moved from Leningrad to Moscow with the rest of the academic institutions in 1934. There are three portraits in the Institute logo, Lomonosov, Kurnakov, and Chugaev. Nikolai S. Kurnakov (1860–1941) graduated from the Mining Institute in St. Petersburg in 1882. He became a physical chemist and one of Mendeleev's disciples. Between 1902 and 1930 he was head of department of general chemistry at the Polytechnic Institute. He founded and directed the Institute of Physical-Chemical Analysis. He directed other institutions as well and by combining them the predecessor of the Kurnakov Institute was formed. Lev A. Chugaev (1873–1922) followed Mendeleev in the chair of chemistry at St. Petersburg University. He was among the pioneers of coordination chemistry.

The Kurnakov Institute has been in the frontiers of learning about the structure of matter. Working in this area under the Soviet conditions was not easy because of political interference. When two of the most advanced researchers of the Institute, Yakov K. Syrkin (1894–1974) and Mirra E. Dyatkina (1915–1972) applied Linus Pauling's theory of resonance for describing the structure of molecules, they were severely criticized (see more about this controversy in Chapter 2). By this time Syrkin and Dyatkina had made their names in the international community of chemists. Their book about structural chemistry, published originally in Russian, was published in English translation in 1950 (*Structure of Molecules and the Chemical Bond*).

One of the novel molecules discovered in the Institute, Re_2X_8, was depicted in front of the image of the Institute on the postage stamp issued for the 50th anniversary of the Kurnakov Institute. X refers to chlorine and the corresponding substance was discovered in the late 1950s by Ada S. Kotelnikova (1927–1990) in this Institute. Her name is hardly known to anybody even within the Institute. She made the discovery but it was not followed up in her homeland. Rather, related work rapidly expanded in the United States where it was determined that this structure contained hitherto unknown strong bonding between two metal atoms. Numerous articles, monographs, and conferences followed, and awards as well, though not for Kotelnikova. Let this paragraph be her memorial.

Frumkin Institute

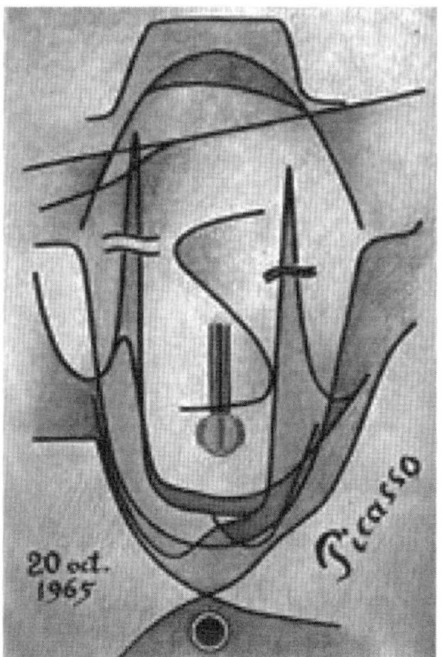

Left: Memorial tablet of Vladimir A. Kistyakovsky on the façade of the Frumkin Institute. *Right*: A collage from Alekszandr Frumkin's research papers in Picasso's style (1965) by Frumkin's pupils. Courtesy of László Kiss.

The Frumkin Institute of Physical Chemistry and Electrochemistry and the Kurnakov Institute share a campus at 31 Leninsky Avenue. The Frumkin Institute was founded in 1945, but its history goes back to the late 1920s when its predecessor laboratories had begun their operations. The physical chemist Vladimir A. Kistyakovsky (1865–1952) was its first director as he took over some of the predecessor institutions in 1934.

The director of another unit of the current Institute was the internationally renowned electrochemist Aleksandr N. Frumkin (1895–1976). He also headed the department of electrochemistry founded by him at Moscow University in 1933. He spent two years in his youth in Strasbourg and Bern to learn physical chemistry and had started publishing papers even before college. He graduated from the Novorossiisky University in Odessa and began his career at the Karpov Institute of Physical Chemistry (Chapter 4) in Moscow. In 1928–1929, he taught at the University of Wisconsin in Madison. In WWII he helped the war efforts with his research in radiochemistry and with his activities in the Jewish Anti-Fascist Committee. In 1949, at the time of the anti-Semitic campaign, he and one of his co-authors, Yakov Syrkin, were accused of under-appreciating the contribution of Russian scientists in the development of physical chemistry. Frumkin was removed from his positions and exiled to a village for a while. During this period, and after he had been reinstated in his positions, he continued his research until the end of his life. For a sustained period, he was one of the most prominent electrochemists internationally.

Metallurgy

Bust of Aleksandr Baikov (P. I. Bondarenko) in front of the Baikov Institute of Metallurgy and Materials Science, 49 Leninsky Avenue, and a memorial plaque of Aleksandr Samarin on its façade.

The Baikov Institute of Metallurgy and Materials Science of the Academy of Sciences was founded in 1938 and its first director, Ivan P. Bardin (1883–1960), was a graduate of the Kiev Polytechnic Institute. He spent two years in his youth in the United States working as laborer in metallurgical plants. In 1932, he was made into an academician, in 1937, he was appointed deputy minister of metallurgy of iron and steel, and in 1942, vice president of the Academy of Sciences. Heavy industry was at the core of the 5-year plans. Today, there is a Bardin Central Institute of Iron Metallurgy (23 Radio Street).

The Institute was named after Aleksandr A. Baikov (1870–1946), a physical chemist and a disciple of Dmitry Konovalov at St. Petersburg University. In time, Baikov followed Konovalov and Konovalov's predecessor, Dmitry Mendeleev, in the chemistry chair. Baikov was also a leading professor at the Leningrad Polytechnic Institute. When the Institute of Metallurgy was formed, Baikov moved to Moscow, was appointed to head its division of metals, and he stayed in this position to the end of his life.

Aleksandr M. Samarin (1902–1970) was a graduate of the Moscow Mining Academy and a specialist of the interactions of gases with steel. He had leading positions in the Institute of Steel and in the Baikov Institute as well as elsewhere, and for a few years he was also deputy minister of higher education.

Machine Science

Blagonravov Institute of Machine Science at 4 Maly Kharitonievsky Street (courtesy of Nadezhda Ismailova).

The initial building of the Blagonravov Institute of Machine Science has elements of English architecture. It was created by contributions of the alumni of the Moscow Institute of Technology (today, Bauman University) to be the seat of the Polytechnical Society. Nikolai Zhukovsky, Petr Lebedev, Konstantin

Tsiolkovsky, and other luminaries of science and technology used to lecture at the meetings of the Society.

The Institute started its operations in 1938 and it has combined fundamental research and applications in direct response to the needs of the national economy and defense. These tasks have changed, from the period of accelerated industrialization to WWII, to meeting the challenges of fast progress in engineering and technology. The areas of activities have included aviation, the conquest of Space, tool engineering, car industry, robotics, defense, and areas of energy production.

Evgeny A. Chudakov (1890–1953; for his memorial plaque, see, p. 152) specialized in internal combustion engines. During his studies, he worked as a mechanic in a factory. Upon graduation, he spent a few years in England at the time of WWI. There was only one plant for the production of automobiles in the Soviet Union at the beginning of the Soviet era and even that plant was merely assembling cars from imported parts. Chudakov initiated a whole new field for automobile production, including theoretical investigations preparing for further development. He suggested organizing research for all aspects of machine building and founded in 1938 the Institute of Machine Science. He continued practical work as well, and introduced numerous innovations in the emerging Soviet automobile industry. He headed the department of automobiles at the Lomonosov Moscow Auto-Tractor Institute. Later, he served as vice president of the Academy of Sciences. He was the principal author of the 15-volume handbook of mechanical engineering published between 1947 and 1950.

From left to right: Memorial plaques of former directors of the Blagonravov Institute, Anatoly A. Blagonravov; Ivan I. Artobolevsky; and Konstantin V. Frolov. Courtesy of Nadezhda Ismailova.

Anatoly A. Blagonravov (1894–1975) was a researcher in mechanics. He studied at the Petrograd Polytechnic Institute and in military academies. He fought in WWI and in the Civil War. His research concerned the higher layers of the atmosphere and he used rocket technology for gathering information. This brought him into close contact with space exploration.

Ivan I. Artobolevsky (1905–1977) was a scientist, engineer, and trade union leader. His father was executed during Stalin's terror in 1938. Artobolevsky graduated from the Timiryazev Academy in 1926 and graduated in mathematics from Moscow University in 1927. He taught at a variety of institutions of higher education, among them, Moscow University, where he co-founded the department of applied mechanics, and was also professor at the Moscow Aviation Institute. He was a life-long associate of the Blagonravov Institute from 1937.

Konstantin V. Frolov (1932–2007) was a specialist in machine science. He was five years old when his father was arrested in 1937 at the time of Stalin's terror. During Frolov's high school studies he held a job as laboratory assistant at his school. He graduated from the Bryansk Institute of Transportation Machines in 1956 and completed his PhD-equivalent studies at the Blagonravov Institute in 1961. He remained at this Institute to the end of his life. He organized new research directions and new divisions, including one in biomechanics. He taught at various institutions of higher education of technology, including Bauman University.

Organic Chemistry

Left: Zelinsky Institute of Organic Chemistry, 47 Leninsky Avenue. *Right*: Memorial plaque of Ivan N. Nazarov (1906–1957) on the façade. Nazarov studied at the Timiryazev Academy and worked at the Zelinsky Institute between 1934 and 1947.

Nikolai D. Zelinsky (1861–1953) was orphaned and brought up by his grandmother. He graduated from Novorossiisky University in 1884. His principal field was petroleum chemistry and catalysis. In 1893 he was already a professor at Moscow University and stayed there through his entire career except for the period 1911–1917. Zelinsky and many of his fellow professors left the University in 1911 in protest against the arbitrary policies of the czarist minister of education. In the mid-1930s, Zelinsky was active in setting up the Institute of Organic Chemistry of the Academy of Sciences. Several major groups joined to form the Institute including the groups of two professors who by then had left the Soviet Union, Vladimir N. Ipatiev and Aleksei E. Chichibabin. Zelinsky led one of the major sections of the Institute until the end of his life and the Institute was named after him upon his death. Many of Zelinsky's disciples became leading scientists in the Soviet Union; some figure in our text, such as Aleksei Balandin, Lev Chugaev, Sergei Nametkin, and Aleksandr Nesmeyanov.

Aleksandr Nesmeyanov and His Institute

Bust of Aleksandr N. Nesmeyanov (N. I. Komov, 1985) in front of the Institute of Element-Organic Compounds (INEOS), 28 Vavilov Street.

Aleksandr N. Nesmeyanov (1899–1980), a brilliant organic chemist, was one of the chief science administrators of his time in the Soviet Union. He graduated from Moscow University in 1922. He was then Nikolai Zelinsky's doctoral student. He started his independent career in 1928 and he fell in love with metal-organic compounds at his first work place in a pesticides laboratory. He formed a group and soon the group developed into a laboratory and this kind of expansion became a repeating pattern for Nesmeyanov. When he was director of the Zelinsky Institute, his own laboratory of element-organic chemistry developed into a separate Institute of Element-Organic Compounds (INEOS). He coined the name *element-organic* — it is *hetero-organic* in the rest of the world. INEOS today carries his name.

Two among the internationally famous associates of INEOS have been Aleksandr I. Kitaigorodsky (1914–1985; Chapter 7) and Elena G. Galpern (1935–). Galpern was the first who calculated the structure of the molecule, known today as *buckminsterfullerene*. Nesmeyanov had suggested constructing molecules in which carbon cages might envelope single atoms or small molecules and Galpern pioneered quantum chemical calculations for such systems. She and her colleagues predicted the shape of the soccer ball for the cage molecule consisting of sixty carbon atoms, C_{60}. She published her findings in a respectable Russian journal in the early 1970s, but her discovery remained unnoticed. Galpern and her work were re-discovered when the molecule had been observed experimentally in 1985, made headlines, and won a Nobel Prize for British and American scientists in 1996.

Memorial plaques of Aleksandr V. Fokin (*left*) and Yury T. Struchkov (*right*). Courtesy of Inga Ronova.

There are a number of memorial plaques at INEOS of which only two are depicted above. Kuzma A. Andrianov (1904–1978) was a chemistry graduate of Moscow University and worked for the aviation industry before he joined the new INEOS. He had clear ideas about the needs of special materials in industry. He organized the laboratory of silicon-organic compounds in 1954, produced silicon-based polymers with properties that made them useful, and developed a school of inorganic polymer chemists.

Ivan L. Knunyants (1906–1990) graduated as chemical engineer from Bauman University in 1928. By then, he was already doing research with Aleksei Chichibabin. He was at the Military Academy of Radiation, Chemical, and Biological Defense (in the town Kostruma) and eventually moved to INEOS. He was not only an academician but also a much decorated general in the armed forces.

Martin I. Kabachnik (1908–1997) graduated from the Mendeleev Institute of Chemical Technology. Before joining INEOS he worked in various organic chemistry laboratories, including one at a military chemical academy. His specialty was phosphorus organic chemistry. He built up a strong laboratory in this field and an internationally renowned school.

Vasily V. Korshak (1909–1988) graduated from the Mendeleev Institute of Chemical Technology. His career was at his Alma Mater and at the Academy of Sciences, first at the Zelinsky Institute, then, at INEOS. In each of these institutes he served as deputy director. He was in charge of a laboratory of polymer chemistry he founded in 1938.

Aleksandr V. Fokin (1912–1998) was a chemical engineer who alternated between civilian and military appointments. From 1945, for two years, he was a member of the Soviet control commission gathering information about the chemical industries of Austria and Hungary. From 1955 he worked at the Military Academy of Chemical Defense and earned his higher scientific degrees there. Between 1960 and 1973 he headed the department of rocket fuels of the Military Academy. He was director of INEOS between 1980 and 1988. He distinguished himself in fluorine chemistry.

Mark E. Volpin (1923–1996) was a graduate of Moscow University. He started his career at the Institute of Scientific and Technical Information. From 1958, he worked at INEOS and headed the laboratory of metal-organic catalysts. He was director of INEOS from 1988 to the end of his life.

Yury T. Struchkov (1926–1995) studied at Moscow University and was a disciple of Aleksandr Kitaigorodsky (Chapter 7) in crystallography. He worked in Kitaigorodsky's laboratory at the Zelinsky Institute and then at

INEOS where in 1973 he continued as head of his own independent unit of crystal structure analysis. Struchkov built up a well-equipped laboratory where analyses were performed on crystals from anywhere in the Soviet Union. He never let out any information from his laboratory without personally checking the data; he was a workaholic. In 1992, he was given an Ig Nobel Prize in Literature for having published 948 papers, an extraordinary output, in the decade of 1981–1990. The Ig Nobel Prize is a parody, but sometimes it awards research that appears initially unnecessary, but may later turn out to be valuable. In Struchkov's case, and we had known him for decades, there was no doubt that he produced much needed information. He was known to be meticulous and conscientious and not someone who would have added his name to the authorship of papers without actual participation in the work.

Returning to Nesmeyanov, he was president of the Academy of Sciences from 1951 until 1961. A robust expansion of the system of research institutes took place under his leadership. One of the twenty-five new research institutes was for scientific information. However, two crucial areas remained in very poor state. One was everything under the umbrella term "cybernetics," which included computer technology. The other was biology.

Biologists and Their Institutes

By the time of Nesmeyanov's presidency, the science of biology in the Soviet Union had sunk deep under the charlatan Trofim Lysenko's domination. World-renowned biologists had perished; and nobody seemed to be able to do anything about it. Nesmeyanov was paralyzed under Stalin to alleviate the situation. When Khrushchev came to power, first there was hope, followed by disappointment. Khrushchev was almost as enamored with Lysenko and his unfounded promises as was Stalin.

Nesmeyanov felt that as Academy president he had to make an effort, and however improbable this sounds, he resorted to conspiracy. He met clandestinely with two leading scientists, Vladimir A. Engelhardt and Mikhail M. Shemyakin, and offered each a new institute with biological flavor. The Academy was about to set up a big new program for chemistry and Nesmeyanov decided to sneak some biology into this new program. One was to be the Institute of Radiation and Physical Chemical Biology with Vladimir Engelhardt as director; today it is the Engelhardt Institute of Molecular Biology in the former building of the Mining Institute (see below).

Sculpture of the institute mascot, the antibiotic valinomycin enveloping a potassium ion, in front of the Shemyakin-Ovchinnikov Institute of Bioorganic Chemistry (courtesy of Ekaterina Altova).

The other was to be the Institute of Natural Products Chemistry with Mikhail Shemyakin as director. Today it is the Shemyakin-Ovchinnikov Institute of Bioorganic Chemistry, which is a center of research of modern biology and biotechnology in Russia. It is in a new location at 16 Miklukho-Maklaya Street.

Memorial plaques of Mikhail M. Shemyakin (*left*) and Yury A. Ovchinnikov (*right*) high on the façade of the Shemyakin-Ovchinnikov Institute (courtesy of Ekaterina Altova).

Mikhail M. Shemyakin (1908–1970) graduated in chemistry from Moscow University in 1930. He worked in various research institutes and developed an interest in the chemistry of biologically active compounds, among them, vitamins, peptides, and amino acids. It was a recognition of his scientific acumen that Nesmeyanov selected him to direct the new Institute of Natural Products Chemistry of the Academy of Sciences in 1959. Shemyakin died prematurely while participating in a conference on natural products. Yury A. Ovchinnikov (1934–1988) was originally one of Shemyakin's disciples, and became his successor as institute director. Ovchinnikov had a meteoric career both in academia and in the communist party. At the age of forty, he was already vice president of the Academy of Sciences. He also rose to leading positions in international organizations. He specialized in the chemical aspects of molecular biology and worked with a large research group. He supported research on chemical and biological weapons. When he died, his name was added to the already existing name of the institute.

As we have seen, Aleksandr Nesmeyanov played a pivotal role in founding the Engelhardt Institute and the Shemyakin-Ovchinnikov Institute. He had a complex personality who had a good deal of knowledge about the horrors of the Soviet regime even at a personal level. In 1941, his brother Vassily fell victim of Stalin's terror and was executed. Yet Nesmeyanov, as many like him, served Stalin and his successors faithfully.

Left: Complex of biological research institutes, 33 Leninsky Avenue. *Right*: Memorial plaque of Aleksei N. Bakh on the façade.

There are several research institutes of the Academy of Sciences, mostly in biological areas, at 33 Leninsky Avenue, among them, the Severtsov

Institute of Ecology and Evolution; the Bakh Institute of Biochemistry; and the Parasitological Institute. Most of these institutes carry the names of their founders. On the façade of the principal building, there are several memorial plaques of which only Bakh's and Oparin's are shown here.

Aleksei N. Bakh (1857–1946) was a founder of biochemistry in the Soviet Union. When he was a student at Kiev University, he participated in an anti-establishment movement. He was exiled without the possibility of continuing his studies anywhere for three years. When he finally resumed his studies, it did not last long as in 1883 he joined an underground revolutionary movement. Another exile followed; this time in France, America, and Switzerland. He started publishing scientific papers of his studies of cell respiration. When in 1917 he returned to Russia, he dedicated himself to organizing science and in 1920 he founded what is today the Bakh Institute of Biochemistry. He was the oldest academician in 1939 when he proposed Stalin's election to honorary membership of the Academy of Sciences.

Another memorial plaque honors Vladimir L. Komarov (1869–1945), a scientist and a science administrator whose main interest was in understanding the flora. He thought that it can only be accomplished by understanding the history of its evolution, the migration of various plant species, and the impact of climate, soil, and the seas. In his research he combined his two major areas of expertise, botany and geography. From early on Komarov was conscious of the importance of the Academy of Sciences not only for science but for the whole country. He proposed to create a national network of research institutes and in particular, the establishment of an independent institute of genetics. He ascribed great importance to studies of the history of science. He was the President of the Academy of Sciences from 1936 until his death. In 1937, he signed a collective letter of scientists condemning Trotsky, Bukharin, and others as traitors and thus helping Stalin in the liquidation of his fellow politicians. In 1941, Komarov chaired a commission that was charged with mobilizing the resources of the Ural Mountains, Western Siberia, and Kazakhstan for satisfying the defense needs of the country.

Yet another memorial plaque is for Konstantin I. Skryabin (1878–1972), a veterinarian. He specialized in parasitological research, the human and animal pathology caused by parasites, and the methodologies of eliminating such illnesses. He is considered the founder of this field in the Soviet Union (more about him in Chapter 5).

Memorial plaque of Aleksandr I. Oparin.

Aleksandr I. Oparin, (1894–1980) earned international fame for his theories about the origin of life at the time when such considerations were yet in their infancy. He graduated from Moscow University and at the age of 30 he was already lecturing about the origin of life. Soon he was giving a course on the chemical foundations of life processes. In 1935 he joined Bakh at his Institute and initiated a laboratory of evolutionary biochemistry and sub-cellular structures. For a long time, Oparin headed the Bakh Institute and the department of plant physiology at Moscow University. He was active in popularizing science and the topic of the origin of life was something that always drew large audiences to his presentations. When the International Society for the Study of the Origin of Life was formed, he was elected its first president.

In the years of Lysenko's domination of Soviet biology, Oparin was Lysenko's strong supporter. Later he explained his behavior by his fear of repression. However, Oparin supported not only Lysenko, but also Lepeshinskaya's unscientific teachings (Chapter 8) and continued doing so when other scientists already had spoken up against these charlatans. Oparin was an original researcher but for the sake of his career, he supported the regime with deeds and words that were to the detriment of fellow scientists that did not sacrifice their principles for personal advancement.

Norair M. Sisakiyan (1907–1966) researched the mechanism of fermentation, the processes of metabolism, the biochemistry of drought-resistant plants, technical biochemistry, and the biology in Space. He worked at the Bakh Institute of Biochemistry and taught at Moscow University.

Vladimir E. Sokolov (1928–1998) was a graduate of Moscow University who became professor of biology at his Alma Mater and the director of the Severtsov Institute of Ecology and Evolution. His main interest was in studying mammals. He had high positions in the Academy of Sciences and often participated in popular science television programs.

Another memorial plaque is for Merkury S. Gilyarov (1912–1985; see also in Chapter 7) who worked in zoology, entomology, evolution theory, and related areas. He studied at Kiev University, started his activities in agriculture, and for years was involved with the protection of plants. In 1936 he moved to Moscow and founded soil zoology in the Soviet Union with two articles published in 1939. He worked both in academy institutes and at the Moscow Pedagogical University.

Nikolai Vavilov

Left: Memorial of Nikolai I. Vavilov at 33 Leninsky Avenue. *Right*: Vavilov's memorial (K. S. Suminov, 1970; courtesy of Galina Bous) in the Voskresenskoe Cemetery in Saratov.

The Genetics Laboratory of the Academy of Sciences was formed in 1930 in Leningrad. It was already an institute when it moved to Moscow to 33 Leninsky Avenue where it stayed until 1940. Nikolai Vavilov was its director until his arrest in 1940. Between 1941 and 1965, Trofim Lysenko was the director of the Institute of Genetics. In 1965 the Institute was dissolved and a new Institute of General Genetics, 3 Gubkin Street, was formed in its stead, which was named after Nikolai I. Vavilov.

Wall of remembrance of Nikolai I. Vavilov in the entrance lobby of the Institute.

Nikolai I. Vavilov (1887–1943) was a biologist, geneticist, agronomist, and explorer. He was also president, then, vice president of the Academy of Agricultural Sciences, president of the Soviet Geographical Society, the founding director of the Institute of Plant Breeding in Leningrad, along with other functions and positions. He was a graduate of the Moscow Agricultural Institute — today, the Timiryazev Academy — where Nikolai Khudyakov and Dmitry Pryanishnikov (about both, see Chapter 6) were among his professors. Nikolai Vavilov had a spectacular career with a tragic ending.

He led expeditions to about fifty countries; initiated a unique collection of seeds; developed a theory of relationships among plants for which he was compared to Linné and Mendeleev; acquired international fame among botanists and geneticists in the world; and lifted the reputation of Soviet science in his field to universal respect and appreciation. Vavilov's fate irrevocably became connected with the unscientific science dictator Trofim D. Lysenko (1898–1976).

Lysenko was an agronomist whose initial interest was in plant breeding to which at the early stage of his career he was devoted unselfishly. He lacked a science education, spoke no foreign languages, did not follow foreign literature, and was not familiar with genetics, especially not with its recent progress. He wanted to shorten the time plants needed to reach harvest and to extend plant growing towards the northern regions of Russia. Both of these goals appeared very attractive for the national economy. Lysenko's devotion made a good impression on his peers and superiors and he kept moving up in scientific circles. Eventually, he caught the attention of the political leaders, ultimately, Stalin's attention.

Statue of Ivan Michurin (D. S. Zhilov, 1958) in the park near the permanent exhibition of the achievements of the Russian national economy (courtesy of Olga Dorofeeva).

Nikolai Vavilov was also taken by Lysenko's enthusiasm and diligence. Vavilov was on the lookout for talent and for ways and persons that could help his field and the economy. He had also "discovered" the innovative plant breeder Ivan V. Michurin (1855–1935). The self-educated Michurin lived in Kozlov (now, Michurinsk) to which Voronezh is the nearest large city, and he developed a magnificent orchard on a rented lot. He bred hybrids of fruit trees by cross-breeding local varieties with far-away trees resistant to cold and drought. His work was based on intuition and many years of experience.

Nikolai Vavilov did more than anybody else to make Michurin's achievements known and help Michurin and his orchard thrive. He freed Michurin from everyday worries and from poverty. Michurin was a great example, but it was a mistake to consider him a great theoretician and create an aura of "Michurinism" as Lysenko and his followers did. By hiding behind Michurin's cult, Lysenko introduced and misused the artificial division between "Michurinists" and "anti-Michurinists."

Lysenko had the ability to paint a picture of a bright future when his innovations would bring enhanced harvests and Vavilov fell for it. He praised Lysenko and the virtually limitless possibilities Lysenko's discoveries had opened up. Vavilov was the first who raised the possibility of electing Lysenko into the Academy of Sciences. The "peasant-scholar" Lysenko was a tailor-made mascot for Stalin's anti-intellectual campaign in the mid-1930s. The "intellectual-scholar" Nikolai Vavilov was an unwitting instrument advancing Stalin's goals. Lysenko found a philosopher supporter in the person of Isai I. Prezent who was ignorant of biology but an expert in careerism. He coined such nonsensical but politically attractive terms as progressive biology, creative Darwinism, Michurinist Darwinism, and suchlike. He found the teachings of the French Jean-Baptist Lamarck applicable to promote Lysenko's techniques. They were about the beneficial effects of changing external conditions on inherited characteristics of plants and animals.

Archival photograph of a sculpture of Stalin and Lysenko.

Lysenko was mastering political skills and becoming increasingly aggressive. When in 1935, serious scientists criticized Lysenko for becoming a dictator in agricultural science, Nikolai Vavilov defended Lysenko. In the same 1935, Lysenko gave a speech in Stalin's presence in which he assigned the character of class struggle to the difference in opinion between supporters and opponents of his "scientific" views. Stalin's public reaction was a "Bravo, Comrade Lysenko, bravo." Within three months, Lysenko was an academician and within three years Lysenko was president of the All-Union Lenin Academy of Agricultural Sciences. It was an ominous sign that Nikolai Vavilov was the previous president, and now he was relegated to the vice president position.

Although Lysenko's techniques failed to result in greater harvests, he was always ready to come up with new proposals that promised spectacular results — never right away but neither in a too distant future. Lysenko and Stalin were in accord. Lysenko advocated the inheritance of acquired characteristics and Stalin wanted to transform people *quickly*. As Lysenko's star was rising, Vavilov's was sinking. The first significant change was that after 1933 Vavilov was never again allowed to travel abroad.

By 1937, more brutal forces came into practice than bona fide scientific disputes. Biologists who were critical of Lysenko were arrested, sentenced, and executed or simply disappeared. The struggle between Lysenko and the scientists finally consumed Vavilov who was declared to be an idealist and a "Morganist-Mendelist" — practically a death sentence. The current American evolutionary biologist and geneticist Thomas Hunt Morgan (1866–1945) and the historic figure Gregor Mendel (1822–1884) were not considered to be mere scientists, but agents of Western imperialism. It is a testimonial to Vavilov's authority that it took a tremendous amount of slander and baseless accusation during an agonizing period of several years before the authorities dared to arrest him in 1940.

Nikolai Vavilov's fate was then sealed; months of brutal interrogation led finally to his confessing to anything the authorities wanted him to confess. He was sentenced to death and he spent months on death row before his sentence was commuted. It no longer mattered; the scientist whose life was devoted to improving his country's food situation was left to starve to death. He died in prison in 1943 in Saratov and buried in an unmarked mass grave.

Many years after he had been exonerated from his "crimes," a group of geneticists decided to erect a memorial to Nikolai Vavilov in the Voskresenskoe Cemetery in Saratov. Vavilov's resting place is still unknown, but for many years victims of political repression were buried there. The official organs to which the scientists had turned for assistance left the petitions unanswered. The necessary funds came together from private contributions and the memorial was ready for unveiling in 1970. When this happened, the mourners were stunned. What happened was that the state authority without whose permission the memorial could not be erected had forced the sculptor to change the face of the statue. Wrinkles had to be removed and a smile had to be carved on the face lest the monument would suggest harsh prison conditions and an unhappy past (see p. 56).

In 1997, another memorial was erected in downtown Saratov on Kirov Square at the beginning of Nikolai Vavilov Street. The occasion was the 110th anniversary of Nikolai Vavilov's birth.

Engelhardt Institute

Left: The Engelhardt Institute of Molecular Biology, 32 Vavilov Street. The building was erected in the years 1951–1954 for the Mining Institute. *Right*: Bust of Vladimir Engelhardt in the entrance lobby (courtesy of Larissa Zasourskaya).

Destroying scientists of Nikolai Vavilov's caliber and so devoted to the Soviet Union was not only a crime; it was also the highest degree of irrationality, and there were plenty of other examples. Thus, expressions such as molecular biology were anathema to the Soviet authorities. It was a long time before the designation of an important institute of the Academy of Sciences could include such a term. Above, we have seen how a clandestine action at the Academy of Sciences helped to establish a modern biology research institute in the abandoned building of the Mining Institute. The façade of the Engelhardt Institute of Molecular Biology still displays four reliefs (not shown here) that were meant to decorate the Mining Institute.

Boris I. Boky (1873–1927) studied at the Mining Institute in St. Petersburg and became a professor at his Alma Mater. He was among those who investigated the 1908 mining catastrophe at Yuzovska in which 274 people were killed and he was very critical of the existing conditions.

After WWI and the revolutions, Boky participated in rebuilding and modernizing the mining industry.

Aleksandr P. Karpinsky (1847–1936) was a geologist who was among a small group of academicians who set out to save the Academy of Sciences in the new order following the Bolshevik takeover in 1917. He became the first president of the Academy who was *elected* to this position. He held the presidency from 1917 to the end of his life. As a student, Karpinsky studied mining engineering in St. Petersburg and became a professor there. His courses included historical geology, petrography, and the occurrence of ores. He conducted broad-based research and was a multifaceted scientist.

Mikhail M. Protodiyakonov (1874–1930) graduated from the Mining Institute in St. Petersburg in 1899. He could not take a job in higher education because he was under police surveillance due to his involvement in the social democratic movement. Later he had professorial appointments in mining institutions. His most important activities were related to lifting the level of professional training in Central Asia where he was among the founders of the oldest institution of higher education there — today the Mirzo Ulugbek National University of Uzbekistan in Tashkent.

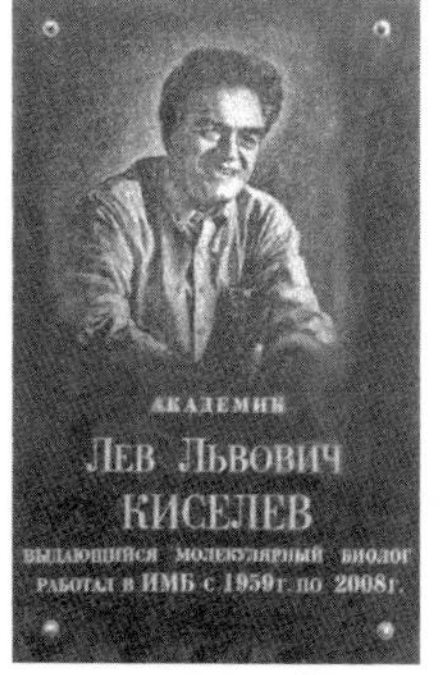

Memorial plaques *from left to right*: Aleksandra A. Prokofieva-Belgovskaya, Lev L. Kiselev, and Andrei D. Mirzabekov (Prokofieva-Belgovskaya and Mirzabekov, courtesy of Larissa Zasourskaya; Kiselev, courtesy of Tatyana Avrutskaya).

In addition to the Engelhardt bust, there are six memorial plaques of other leading molecular biologists in the entrance lobby of the Institute, three of which are shown here. Vladimir A. Engelhardt (1894–1984) graduated from the Faculty of Medicine of Moscow University in 1919. Nikolai K. Koltsov especially influenced his scientific interests. Engelhardt served as medic in the civil war, 1919–1921. He began his scientific career in the Institute of Biochemistry, which included a few months of learning in

Peter Rona's laboratory of medicinal chemistry in Berlin in 1927. Engelhardt worked as professor of biochemistry and as head of department in parallel and in succession at Kazan University, Leningrad University, the Bakh Institute of Biochemistry, the Pavlov Institute of Physiology in Leningrad, and the Institute of Experimental Medicine. He became director of the new Institute of Radiation and Physical Chemical Biology in 1959 under the circumstances mentioned above.

Aleksandr E. Braunshtein (1902–1986) biochemist and physician by training worked much of his career in interaction with Engelhardt. Braunstein's main interest was in amino acids and in their transformations and structures. He was a foreign member of the US National Academy of Sciences.

Aleksandra A. Prokofieva-Belgovskaya (1903–1984) geneticist lost her job and the possibility of doing research in 1948 as a consequence of Stalin and Lysenko's anti-science campaign. Engelhardt helped her return to her original research and she became a pioneer in human cytogenetics.

Aleksandr A. Baev (1904–1994) biochemist and physician by training began his career with Engelhardt in Kazan and worked in the Engelhardt Institute in Moscow during the last decades of his life. In between, he was twice incarcerated for years on trumped up charges and worked as a physician in labor camps. His main interest was in bioenergetics and bio-polymers, and he was in charge of the human genome program in the Soviet Union.

Aleksandr A. Kraevsky (1932–1999) was a bioorganic chemist who investigated potential drugs against the acquired immune deficiency syndrome (AIDS).

Lev L. Kiselev (1936–2008) was a son of the immunologist and virologist Lev A. Zilber. Zilber's other son was also a virologist, Fedor L. Kiselev (1940–2016). About Zilber's life and the reason why father and his sons had different surnames, see Chapter 5. Lev Kiselev graduated from the Faculty of Biology and Soil Science of Moscow University in 1959. He was not allowed to stay for post-graduate studies, because as a sophomore, he attended a student circle for studying genetics. In the mid-1950s, Lysenko was still the supreme authority in Soviet biology. However, Engelhardt's Institute took him in for post-graduate research. This Institute remained Kiselev's work place for the rest of his life. He was among the founders and the Editor-in-Chief of the Russian periodical *Molecular Biology*. Between 1992 and 2005 Kiselev chaired the scientific council of the Russian program "Human Genome."

Andrei D. Mirzabekov (1937–2003) studied at the Lomonosov Institute of Fine Chemical Technologies and already as a student he worked in Engelhardt's Institute. He became director of the Institute in 1984.

Mirzabekov's biographies note that his stays as visitor at leading western laboratories contributed to his becoming a renowned molecular biologist. We may add that Mirzabekov also contributed to his western colleagues' discoveries in a creative way. A case in point was when in 1975 he visited the physicist-turned-biologist Walter Gilbert at Harvard University, and Mirzabekov suggested to Gilbert a crucial experiment. Gilbert was reluctant to follow Mirzabekov's advice, but Mirzabekov happened to be on yet another visit and finally convinced Gilbert to do the experiment. According to Gilbert, it turned out to be an important step in creating a new method for the determination of the base sequence in nucleic acids. This is what already in 1980 brought Gilbert a share of the Nobel Prize in Chemistry.[8]

Koltsov Institute

Left:Nikolai K. Koltsov's bust at the Koltsov Institute. *Right*: Timofeev-Ressovsky's bust at the Timofeev-Ressovsky's House in the molecular biology research center in Berlin-Buch.

The Koltsov Institute of Developmental Biology is at 26 Vavilov Street. Nikolai K. Koltsov (1872–1940) graduated from Moscow University in 1895 majoring in zoology under the mentorship of Mikhail A. Menzbir at the department of comparative anatomy. Koltsov spent various periods in West-European laboratories while he was building his career at Moscow University. He participated in protest movements and this led to his expulsion from

[8] Istvan Hargittai, *Candid Science II: Conversations with Famous Biomedical Scientists*, ed. Magdolna Hargittai (London: Imperial College Press, 2002), p. 104.

Moscow University. He continued at the Higher Courses for Women and at Shanyavsky University. After the 1917 revolution, he returned to Moscow University and stayed there until 1929. Already in 1917 a new Institute of Experimental Biology was founded and Koltsov became its head. It was the first research organization in Russia outside the network of the institutions of higher education.

Today, it is the Koltsov Institute of Developmental Biology, which was formed by joining several research institutions, including one that was directed by Boris L. Astaurov (1904–1974). Astaurov was a biologist graduate of Moscow University and also a Koltsov disciple. Between 1966 and 1972, he chaired the Nikolai I. Vavilov Society, which helped the revival of the science of genetics in the Soviet Union.

Koltsov developed a large and influential school of leading biologists. However his career and creative activities were hindered by various difficulties. Already in 1920 he was arrested on false charges. First he was sentenced to death; then the sentence was commuted; finally, he was freed probably due to the testimonials of famous intellectuals. He was again persecuted following Nikolai Vavilov's arrest in 1940 and as he was being subjected to interrogations, he suffered a fatal heart attack. Upon his death, his wife committed suicide.

Nikolai V. Timofeev-Ressovsky (1900–1981) was among the noted biologists who started their careers as Koltsov's disciples. Timofeev-Ressovsky also learned from Sergei S. Chetvernikov. In the mid-1920s, the Soviet Union was still open for international scientific exchange. According to a Soviet-German agreement, the Germans set up brain research in Moscow and the Soviets helped Germany in genetics. It is remarkable that genetics was at such a level in the mid-1920s in the Soviet Union that the Germans were willing to learn from their Soviet colleagues. At Koltsov's suggestion, Timofeev-Ressovsky went to Berlin for a year or two, but he stayed for twenty years, doing research in the academic center of Berlin-Buch. Timofeev-Ressovsky became an internationally renowned geneticist while the science of genetics was being destroyed in his home country. When he returned to the Soviet Union in 1945, he was treated as a prisoner, yet appointed head of a biophysical laboratory. His expertise in the biological effects of radiation was much in demand due to the nuclear program. To the end of his life, he was barred from living in Moscow.

Statue of Mikhail V. Lomonosov (N. V. Tomsky, 1953) in front of the Main Building of Lomonosov Moscow State University, in between the Faculty of Physics and the Faculty of Chemistry.

2

Lomonosov University

The Lomonosov Moscow State University (or Moscow University)[1] is the flagship institution of higher education in Russia as it was in the Soviet Union. An expression of its high standing was that other top universities in the member republics used to send some of their top students for their senior year to study at Moscow University. It still receives students from some of the former Soviet republics. The founding of this University was an expression of the best ambitions for lifting the level of science and of progress in general in Russia. Empress Elizabeth signed the order establishing Moscow University in 1755 at the initiative of Ivan I. Shuvalov (see, below). The order declared that the new institution was for all to indulge in science regardless of social stratum or whether they do or do not belong to the nobility.

Mikhail Lomonosov

Mikhail V. Lomonosov (1711–1765) was a natural scientist, a polymath, a member of the Russian Academy of Sciences and an honorary member of the Royal Swedish Academy of Sciences. He helped design the plans for Moscow University. He came from the region of Arkhangelsk, 740 km (460 miles) northeast of St. Petersburg, from a well-to-do family of farmers. The name of the village of his birthplace is now Lomonosov after its famous scion. He was looking for something better than what the village life could offer and moved to Moscow at the age of nineteen. He could enroll in the school he wanted to attend only if he pretended to come from a noble family, so he did. He had a strong determination to study and a thirst for knowledge. For a short while he then continued his studies in Kiev. Then, he moved to St. Petersburg to attend a school run by the Academy of Sciences, with emphasis on the natural sciences and German. In 1736, the president of the Academy sent the 12 most gifted students for a study trip to learn more about the sciences and also metallurgy and mining. Lomonosov was among them and he spent some time in Germany and Holland. In addition to the sciences and German, he learned French and

[1] Until 1917, Imperial Moscow University; from 1917, Moscow State University; it was named after Mikhail V. Lomonosov in 1940.

Italian, dancing, drawing, and fencing, and started collecting books. He returned to St. Petersburg in 1741, and his German wife and children joined him two years later. Lomonosov became associated with the Academy in 1742 and was appointed professor of chemistry in 1745.

In 1749, on the occasion of a festive meeting of the Academy, Lomonosov made a rousing speech glorifying Empress Elizabeth II. This was the beginning of his increasing involvement in the Court. In 1753, Elizabeth II bestowed a substantial and inheritable estate upon Lomonosov in Western Russia where he developed a plant to produce mosaics. In 1755, Lomonosov participated in establishing Moscow University. The Russian national poet Aleksandr S. Pushkin called him "Our First University," referring to Lomonosov's versatility and the depths of his knowledge on a universal scale.

Old Campus

The first venue of the new university was the former city hall on the Red Square. It was then the Principal Medicine Store and a few of its rooms served the university until 1784. The building no longer exists. In its place, the venue for the State Historical Museum was erected between 1875 and 1881.

The small building in the center used to house the Laboratory of Medicinal Chemistry and the memorial plaque on its façade remembers Vladimir S. Gulevich.

The Laboratory of Medicinal Chemistry used to be at 2 Bolshaya Nikitsaya Street. Vladimir S. Gulevich (1867–1933) was a graduate of the Medical Faculty of Moscow University and worked here in biological chemistry from 1901 until the end of his life. He was a professor, for a short time rector, and his research included proteins, nucleic acids, and vitamins. In addition to Moscow University he taught at the Higher Courses for Women.

Zoological Museum, 6 Bolshaya Nikitskaya Street (K. M. Bykovsky was the architect).

The Zoological Museum of the University was established in 1791. It has close to five million specimens and receives about a hundred and fifty thousand visitors annually. Its tasks include education and research and assisting the preservation of nature's values through the specimens in its collection. This Museum did something more in the years of the Soviet reign. The Soviet Union paid little attention to the protection of the environment, and the movement of nature preservation was considered anti-Soviet. The center of this movement was the Zoological Museum and it has been called an "island of freedom," which was unrecognized even by most of the keen western observers of Soviet society.[2] Scientists who wanted to save biology against Lysenko supported the idea of establishing nature preserves. Petr Kapitsa was such a scientist and so were Aleksandr Nesmeyanov, Vladimir Engelhardt, and Nikolai Zelinsky. Others, notably, Olga Lepeshinskaya and Aleksandr Oparin, were in their opposition.

Left: 11 Mokhovaya Street is the "old" building of the old campus, built between 1782 and 1793 (design by Matvei Kazakov). It was rebuilt by Domenico Giliardi after the 1812 fire. It used to house the physical-mathematical school. *Right*: Memorial plaque of Petr N. Lebedev who worked in this building. Vladimir I. Vernadsky has a plain memorial tablet (not shown here) on the same façade.

[2] Douglas R. Weiner, *A Little Corner of Freedom: Russian Nature Protection from Stalin to Gorbachev* (Berkeley, CA: University of California Press, 1999), p. 5.

Empress Catherine the Great had the University moved to its new venue at 11 Mokhovaya Street. Initially, there were three faculties: philosophy, medicine, and law. The medical school moved to a new campus in Deviche Pole (Chapter 5) in the last years of the nineteenth century.

Statues of Aleksandr Herzen (*left*) and Nikolai Ogarev (*right*, both by N. Andreev, 1922) in the garden at 11 Mokhovaya Street.

Aleksandr I. Herzen (1812–1870) was a philosopher and writer as well as a revolutionary. He studied at the physical-mathematical section of the University and was increasingly engaged with new ideas of social progress. Early on he was arrested and exiled for years. When he returned and continued his activities, soon, he had to immigrate to Western Europe. He stayed active in revolutionary movements and published books and a newspaper. Nikolai P. Ogarev (1813–1877) was also a journalist and writer as well as a revolutionary. He was Herzen's closest friend. There is a memorial of Herzen and Ogarev on the Vorobyovy Hills where they took an oath to fight the tsarist government.

During the last decades of the tsarist regime there was a growing animosity between the students and professors, on the one hand, and the authorities, on the other. In particular, the minister of education, Lev Kasso, bluntly violated the autonomy of the University. In 1911, many of the professors left the University in protest. Just to mention a few names, Sergei Chaplygin, Petr Lebedev, and Nikolai Zelinsky were among them. After the 1917 revolution, the Soviets took over the University. They expanded the circle of students to include broad layers of society that until then could not have dreamed of attending this School. As for university autonomy, it disappeared completely soon enough. The Communist Party exercised total control for decades, until the 1990 political changes.

Memorial plaque (2008) of Ivan A. Ilyin on the façade of 11 Mokhovaya Street.

Ivan A. Ilyin (1883–1954) was a student and teacher of Moscow University, jurist, writer, rights advocate, and thinker. In 1922, Lenin ordered 160 renowned philosophers, historians, and economists to leave the Soviet Union, including Ilyin. It was the famous Philosophical Steamship taking these intellectuals to foreign destinations. In retrospect, these exiles were the lucky ones. Later, many others were murdered without bothering to exile them. Ilyin had a professorial appointment in a Russian research institute in Berlin from 1923. After the accession of the Nazis to power, Ilyin moved to Switzerland. He left his rich archives to Moscow University.

The "new" building of the old campus at 9 Mokhovaya Street (V. I. Bazhenov and E. D. Tyurin, 1837). Lomonosov's statue in its front (I. I. Kozlovsky, 1957) is the third in this place. The first (S. I. Ivanov, 1876) was destroyed in WWII. The second (S. D. Merkulov, 1945) was made of plaster.

Heroism

Eternal Glory to Fighters Heroes! This memorial (2010) honors the students, recent graduates, post-graduate students, and associates of the Faculty of Physics who were killed in WWII.

WWII demanded tremendous sacrifices from Moscow University as an institution and from many individuals. Memorials to the heroes appear everywhere and we chose to present the one in front of the Faculty of Physics. The majority of the fallen heroes were students. They were of the age of conscription. The relatively small number of faculty associates had a specific explanation. Germany attacked the Soviet Union on June 22, 1941. On September 15, 1941, the State Committee of Defense — the supreme authority during the war under Stalin's leadership — issued an order, which forbade dispatching scientific researchers of research institutions as well as members of the teaching staff of institutions of higher education to the front.

New Campus

Soon after the war, Stalin had grandiose plans that only a totalitarian state could accomplish under the conditions of a war-ravaged country. Such was the secret development of the atomic bomb, which many in America found hard to believe the Soviets would be capable of doing. On a smaller scale, but in front of everybody's eyes, was the construction of seven skyscrapers in Moscow. Their functions were assigned after the projects were already under way. Two of the seven became hotels, two, ministries and offices — one them the Ministry of Foreign Affairs — two, apartment buildings for the Soviet elite, and one was to become the new Moscow University.

The commission for the new building on Vorobyovy Hills — renamed Lenin Hills — went initially to Boris Iofan. He had thought of placing the building at the edge of Vorobyovy Hills, which would have provided a spectacular view of the city below, but there was danger of landslide as well. Finally, Lev Rudnev was appointed to be in charge along with the participation of other renowned architects and hundreds of professional designers. Rudnev moved the planned building 800 meters (half a mile) away from the edge. Thousands of (*GULAG*) slave laborers worked on the construction. The building was inaugurated on September 1, 1953. Aleksandr N. Nesmeyanov spoke at the opening ceremony of the Main Building of Moscow University. Nesmeyanov was rector of the University from 1948 until 1951. Now he was the top science administrator of the country, President of the Soviet Academy of Sciences.

University Tower

View of the university tower and its image in the reflecting pool from the north-east. This is the side of the main entrance used mainly for ceremonial purposes.

Statues of a young man and a young woman with paraphernalia of victory at the top of the Main Entrance.

View of the university tower from the south-west. This is the side of the students' club. There is Lomonosov's statue in the forefront.

Vera I. Mukhina's student statues at the side of the students' club.

The external appearance of the university tower is solemn with all the paraphernalia symbolizing learning, the dominance of workers and peasants, the equality of men and women, and the almightiness of socialism and the Soviet Union.

Two memorial plaques on the façade of the students' club entrance. *Left*: Mstyslav V. Keldysh (Chapter 4) studied and worked here. *Right*: Ivan G. Petrovsky (I. M. Rukavishnikov) was rector for a long time.

Ivan G. Petrovsky (1901–1973) studied mathematics which was stretched out due to the Civil War. He graduated from the University in 1927. He spent his entire career at his Alma Mater; already he was professor at the age of 32. He was appointed head of department of differential equations as well as rector in 1951 and stayed in these positions to the end of his life.

The interior of the University conveys the impression of old-time luxury, but over six-decades of neglect of adequate maintenance has left its marks. For small-scale important events there is the elegant Rotunda at the top floors of the tower (Chapter 3). For large-scale events, there is the Ceremonial Hall, which is often the venue of affairs that may not necessarily be connected with the University, such as international press conferences. The mural of the podium is full of Soviet-era symbolism.

There are two entrance lobbies for the Ceremonial Hall, each with two large statues of world-renowned Russian scientists; Mendeleev and Zhukovsky in one and Michurin and Pavlov in the other. All four were created by Matvei G. Manizer and Elena A. Yanson-Manizer.

Statues of Dmitry I. Mendeleev (*left*) and Nikolai E. Zhukovsky (*right*) in one of the two entrance lobbies of the Ceremonial Hall. More about Zhukovsky in Chapter 4.

Dmitry I. Mendeleev (1834–1907) is one of the most famous scientists of all time. His periodic table of the elements hangs in most classrooms of chemistry everywhere in the world and every introductory chemistry text displays it. He compiled its first version in 1869 while preparing his general chemistry lectures as professor of chemistry at St. Petersburg University. He was not the only one who noticed periodicity in the properties of the elements, but he alone made predictions for yet missing elements on the basis of the regularities he had observed. His predictions have proved correct.

Considering the enormity of importance of his discovery, Mendeleev should have received the Nobel Prize in Chemistry for which he had been nominated repeatedly. There was strong support for his award at the Royal Swedish Academy of Sciences, but the final vote in 1906 went for someone else. This was the last chance for Mendeleev's award as he died in 1907 and there is no posthumous Nobel Prize. Mendeleev was a corresponding member of the Russian Academy of Sciences, elected in 1877. He never made it to full membership; the Academy voted about it in 1880 and declined it. Mendeleev received distinctions internationally; he was elected foreign member of the Royal Society and received its highest award, the Copley Medal, in 1905. Element 101, Mendelevium, carries his name.

Statues of Ivan V. Michurin (*left*, Chapter 1) and Ivan P. Pavlov (*right*) in one of the two entrance lobbies of the Ceremonial Hall.

Ivan P. Pavlov (1849–1936) received the Nobel Prize in Physiology or Medicine in 1904 for his discoveries in the investigation of the physiology of digestion. He began studying for a career in the Church, but having read a paper by the physiologist Ivan Sechenov enticed him to pursue medical studies. All his life he worked in physiology, but he had also qualified as a surgeon. For some time he was the head of the physiological laboratory at S. P. Botkin's clinic. From 1890, Pavlov headed the department of physiology at the Institute of Experimental Medicine in St. Petersburg/Leningrad. After the 1917 revolutions, the Soviet Government supported him, but Pavlov often criticized the brutality of the Soviet regime.

Pavlov achieved most of his major discoveries in studying the physiology of digestion during the last decade of the 19th century. He formulated his teachings about the physiology of the higher nervous activities, including the role of unconditional and conditional reflexes based on his famous experiments with dogs.

After Pavlov's death the Soviet authorities added an ideological aspect to his scientific legacy. They maintained that Pavlov's discoveries were closely related to the ideological bases of the Soviet regime. This had tragic consequences, especially during the early 1950s. Under the pretext of protecting Pavlov's legacy, the Soviet authorities persecuted leading physiologists as part of their anti-science and anti-Semitic campaign during Stalin's last years. All this could not have been farther from Pavlov's teachings and demeanor.

On the walls of the two entrance lobbies, there are scores of mosaic portraits of Russian and international scientists of which a selection, about one third of the total, is shown here.

From left to right and from top to bottom: Archimedes, Butlerov, Copernicus, Curie M, Dalton, Darwin, Descartes, Euler, Faraday, Galilei, Gauss, Kepler, Lobachevsky, Lomonosov, Maxwell, Newton, Pasteur, Pavlov, Rutherford, Sechenov, Stoletov, Timiryazev, Tsiolkovsky, and Vernadsky.

As international relations kept easing from the late 1950s, many outstanding international scientific meetings took place at Moscow University. Often the Ceremonial Hall served as the venue for the opening and closing ceremonies and the plenary lectures. Moscow University was a recognized institution and its physical plant was suitable for holding such gatherings at the time when appropriate venues were scarce in Moscow. International scientists were eager to attend these meetings not only for their scientific values but also for gaining first-hand experience in this isolated super-power and for seeing Soviet colleagues that otherwise were not possible to meet.

The 5th International Congress of Biochemistry was such a meeting in 1961. Its significance was enhanced by the circumstances in that although Lysenko was still a dominant factor in the biological and agricultural sciences, his iron grip was loosening. The international organizers of the meeting stipulated, and were granted by the Soviet hosts, that there should be no interference in their choosing the topics and speakers. There was though a limit to this freedom. It was still prohibited to use the term "molecular biology," which was anathema to official Soviet science. The organizers accepted this limitation and substituted the term molecular biology by expressions such as biological structure and function at the molecular level.

That this meeting could take place was an indication of progress, but apart from that nobody was expecting any major breakthrough to be announced. The plenary lectures went on smoothly and with no sensational revelation in the Ceremonial Hall. Something dramatic though did happen in one of the small rooms where in front of a handful of biochemists a milestone discovery was announced. Marshall Nirenberg of the US National Institutes of Health of Bethesda, Maryland, reported that his experiments together with Heinrich Matthaei resulted in the first step in deciphering the genetic code. In an unprecedented move, the organizers asked Nirenberg to repeat his communication in a plenary presentation arranged impromptu. Nirenberg's talk had an electrifying effect on the participants of the meeting and induced an avalanche of experiments worldwide.

Promenade of Scientists

View from the balcony of the 32nd floor of the university tower in northeastern direction. In the immediate front, there is the reflecting pool with the Promenade of Scientists on its two sides. Further, there is the Moscow River, the Luzhniki sports arena, and downtown Moscow.

There is a long walk from the main entrance of the university tower to the edge of the Vorobyovy Hills. From that edge a spectacular view opens to the river, the Luzhniki sports arena, and further to downtown. It is a favorite spot for newlyweds to visit on the day of their wedding.[3] From the university tower to the edge of the hill and further down to the river, it is all a big park. This park is intersected by Universitetsky Avenue running parallel to the river and to the long axis of the university tower.

There is a reflecting pool between the university tower and Universitetsky Avenue and two rows of busts of scientists, one row on each side of the pool. The busts on the *west side*, starting from the one closest to the tower and progressing toward Universitetsky Avenue are as follows: Pavlov, Michurin, Popov, Dokuchaev, Chernyshevsky, and Lobachevsky (discussed here are only those that are not mentioned elsewhere).

[3]This is one of the two such spots; the other is the Eternal Flame honoring the unknown hero of WWII at the Aleksandrovsky Garden at the Kremlin.

Busts of Ivan P. Pavlov (M. G. Manizer, see above), Ivan V. Michurin (M. G. Manizer, see Chapter 1), and Aleksandr S. Popov (M. T. Litovchenko).

Aleksandr S. Popov (1859–1906) was a physicist, electrical engineer, and inventor; a pioneer of radio in Russia. He was a graduate of St. Petersburg University and taught at a naval academy. He had various inventions utilizing electrical devices and proposed new devices, such as his lightning detector. He demonstrated his radio receiver in 1895, but did not patent it. His inventions in the area of the radio were contemporaneous with the inventions of others, notably Guglielmo Marconi who, however, patented his inventions.

Busts of Vasily V. Dokuchaev (M. G. Manizer, see Chapter 3), Nikolai G. Chernyshevsky (G. V. Neroda), and Nikolai I. Lobachevsky (N. V. Dydykin).

Nikolai G. Chernyshevsky (1828–1889) was a philosopher and a revolutionary socialist.

Nikolai I. Lobachevsky (1792–1856) was a mathematician and geometer, a pioneer of non-Euclidean geometry. He graduated in physics and mathematics from Kazan University in 1811. He stayed on at his Alma Mater; in 1822 he was full professor already, and in 1827, rector. Kazan University dismissed him in 1846, and he died blind and in poverty in 1856. Lobachevsky was not alone in developing a non-Euclidean geometry; the Hungarian mathematician János Bolyai (1802–1860) did that too. The two worked independently from each other. Lobachevsky's non-Euclidean geometry is often referred to as Lobachevskian geometry or hyperbolic geometry.

The busts on the ***east side***, starting from the one closest to the tower and progressing toward Universitetsky Avenue are as follows: Zhukovsky, Timiryazev, Mendeleev, Chebyshev, Herzen, and Lomonosov (discussed here are only those that are not mentioned elsewhere).

Busts of Nikolai E. Zhukovsky (M. G. Manizer, see Chapter 4), Kliment A. Timiryazev (S. D. Merkurov, see Chapter 6), and Dmitry I. Mendeleev (see above).

Busts of Pafnuty L. Chebyshev (I. A. Rabinovich), Aleksandr I. Herzen (see above), and Mikhail V. Lomonosov (I. I. Kozlovsky, see above).

Pafnuty L. Chebyshev (1821–1894) was a mathematician who graduated from Moscow University in 1841. From 1847 he was a member of the professorial staff of St. Petersburg University. In 1852 he visited Western Europe and the visit enhanced his interest in mechanical machines. He is considered to be one of the greatest mathematicians of the 19th century, especially in the theory of numbers, theory of probability, and mechanics. He was a member of 25 academies of sciences.

Library

There are many libraries scattered around the old and new campuses and in 2005, a new central library opened. The two dates, 1755 and 2005, on its main façade refer to the founding of Moscow University and to the opening of the new library.

Bird's eye view of the new Library from the balcony of the 32nd floor of the university tower.

An underground passage leads from the square on which Lomonosov's statue stands to the Library. There are reliefs displaying various structures of the University on the walls of this underground passage as if illustrating a time travel through the history of the University.

Statue of Ivan I. Shuvalov (Zurab K. Tsereteli, 2004) in front of the Library, called officially the Intellectual Center — Fundamental Library, with the founding document of the University in his hand.

Ivan I. Shuvalov (1727–1797) was an influential statesman in the court of Elizabeth II. He founded Moscow University and the Russian Academy of Arts.

Faculties

Today, there are over forty faculties, including institutions that have different names, but have the status of faculty, for example, the School of Business Administration, the Moscow School of Economics, and the School of Contemporary Social Sciences. We mention memorials from only a few of them. The Earth Science Museum is presented in Chapter 3. The central section of the university tower houses the Rector's Office, the Faculty of Mechanics and Mathematics, the Faculty of Geology, the Faculty of Geography, the Earth Science Museum, and others.

Mathematics

The Faculty of Mechanics and Mathematics occupies the floors 12–16. It was formed in the early 1930s; before that, mathematics and the science branches were in the Faculty of Physical and Mathematical Sciences. Tens of thousands of specialists have graduated from the Faculty of Mechanics and Mathematics. Many have become professors at institutions of higher education and distinguished contributors to the nuclear projects and the space program. Until the end of the Soviet regime the Faculty practiced discrimination in that it accepted hardly any Jewish students.

There are memorial plaques of famous mathematicians on the walls of the Faculty of Mechanics and Mathematics of which only a few are presented here. A modest memorial tablet honors Nikolai N. Luzin (1883–1950) next to the auditorium named after him. He founded a strong school and his informal circle had a well-known nickname, "Luzitania." Many of the leading Soviet mathematicians had been his disciples. In 1936, Luzin, a member of the Academy, seemed to be destined to fall victim of the Great Terror. Newspaper articles attacked him and a commission of the Academy endorsed the accusations. Sadly, some of Luzin's former pupils, P. S. Aleksandrov, A. N. Kolmogorov, and A. Ya. Khinchin, among them, turned against their mentor. Renowned scientists, such as S. N. Bernshtein, A. N. Krylov, N. S. Kurnakov, and V. I. Vernadsky among them, defended Luzin. One of his "crimes" was that he published his best results in international periodicals. Petr Kapitsa, who had recently been forced to stay in the Soviet Union (Chapter 1) and was building up his new life, felt obliged to stand up for a fellow scientist. In a letter to the head of government, Vyacheslav Molotov, he explained that it was in the national interest that scientists publish their best papers in international journals. The Soviet authorities continued preventing international publication, but let Luzin off the hook. The Soviet Academy in a 1936 resolution condemned Luzin, but allowed him stay a member of the Academy. He lost his job, but survived. In a 2012 resolution, the Russian Academy of Sciences annulled the 1936 condemnation and finally put an end to the infamous "Luzin Case."

Memorial plaques of Pavel S. Aleksandrov (*left*) and Andrei N. Kolmogorov (*right*) on the façade of Moscow University (courtesy of Olga Dorofeeva).

Pavel S. Aleksandrov (1896–1982) graduated in mathematics from Moscow University. In the early 1920s, he visited Germany and developed joint work with such greats as David Hilbert, Richard Courant, and especially Emmy Noether. For a long time, Aleksandrov chaired the Moscow Mathematical Society. His immediate field was topology and he chaired the department of geometry and topology from its formation in 1933. He was also in charge of the division of topology at the Steklov Institute (Chapter 1) between 1935 and 1950. He was a member of several renowned international learned societies, including the US National Academy of Sciences and between 1958 and 1962, he was vice president of the International Mathematical Union.

Andrei N. Kolmogorov (1903–1987) worked in many areas of mathematics, especially in the theory of probability. He visited David Hilbert and Richard Courant in 1930. Kolmogorov pioneered mathematical linguistics in the Soviet Union. He was appointed professor at Moscow University in 1931 and full member of the Academy of Sciences in 1935 soon after the scientific degree system had been introduced. He founded the department of the theory of probability at Moscow University. Towards the end of his life he founded and headed yet another department, mathematical statistics. Kolmogorov worked for the artillery in World War II and returned to his mathematical research after the war. He helped reform high school math teaching and developed a strong school and some of his disciples continued their careers in the West following the collapse of the Soviet Union. He was a foreign member of the Royal Society (London), the US National Academy of Sciences and other learned societies. In 2002, the University of London established the annual Kolmogorov Lecture and Medal.

Memorial plaques on the façade of the Faculty of Computational Mathematics and Cybernetics: Andrei N. Tikhonov (*left*) and Lev S. Pontryagin (*right*).

Andrei N. Tikhonov (1906–1993) graduated from Moscow University majoring in mathematics and stayed on at the University. In 1933, the Faculty of Physics and Mathematics was divided into the Faculty of Mechanics and Mathematics and the Faculty of Physics. Tikhonov continued at the department of mathematics of the Faculty of Physics, serving as its head between 1938 and 1970. He also held leading positions at three institutions of the Academy of Sciences. In 1970, Tikhonov opened the Faculty of Computational Mathematics and Cybernetics and served as its first dean. There had been a long way from the condemnation of cybernetics as an imperialist conspiracy alien to Marxism-Leninism of around 1950 to this new institution.

In 1955, Tikhonov signed the so-called "letter of three-hundred," in which biologists and other scientists described the state of affairs of biology in the Soviet Union and criticized Lysenko's activities. So many leading scientists signed this letter, including some involved with defense projects, that there was no real danger that the then Soviet leader Khrushchev would strike back.

In 1983, Tikhonov signed another letter and of a very different character. It was a condemnation of Andrei D. Sakharov's activities, accusing him of having lost his honor and conscience and of treason. Only four persons, all of them academicians, signed this letter. Beside Tikhonov, they were A. A. Dorodnitsyn, A. M. Prokhorov, and G. K. Skryabin (they are presented elsewhere in this book). At this time, Sakharov was in exile and living under constant harassment by the KGB.

Lev S. Pontryagin (1908–1988) lost his eyesight at the age of 14 when an oil-stove exploded and badly injured his face. His mother became his eyes and this is how he completed his education. She even learned German so that she could read German papers to him. Pontryagin graduated from Moscow University in 1929 majoring in mathematics. He stayed at the University and he was appointed professor in 1935. In the meantime, from 1934, he was also a leading associate at the Steklov Institute of Mathematics.

From early on Pontryagin was very successful in the applications of mathematics for practical problems. There was a dark side though of his career. He was a dedicated anti-Semite and was active in preventing even the most outstanding Jewish mathematicians from attending international gatherings and getting elected to the Academy of Sciences. In his actions he was helped by his high positions in the hierarchy of mathematicians; by other high-positioned anti-Semites, such as the director of the Steklov Institute, Ivan Vinogradov; and by the general anti-Semitic policies of the Soviet State and the Communist Party.

Faculty of Physics

Central section of the Faculty of Physics (its bird's eye view is presented in Chapter 3).

Moscow University established a separate Faculty of Physics and Mathematics in 1850. In 1903, a Physical Institute was organized at Nikolai A. Umov's (see below) initiative. He and Petr Lebedev were among the professors who left the University in 1911 as a protest against violations of university autonomy.

Between the two world wars — this was already the Soviet era — the Faculty of Physics became very strong with such names among its professors as Nikolai Bogolyubov, Ilya Frank, Dmitry Ivanenko, Grigory Landsberg, Mikhail Leontovich, Leonid Mandelshtam, Otto Shmidt, Pavel Sternberg, Dmitry Skobeltsyn, Igor Tamm, Sergei Vavilov, Vladimir Veksler, and others. A Museum of the Faculty of Physics has operated since 1996. It is not only a depository of memorials, but also a popular venue for study among the students.

Altogether seven Nobel laureates in physics studied and/or taught at the Faculty of Physics, viz., Aleksei Abrikosov, Vitaly Ginzburg, Ilya Frank, Petr Kapitsa, Lev Landau, Aleksandr Prokhorov, and Igor Tamm. The Nobel Peace Prize winner Andrei Sakharov can be added to this list. Having eight Nobel laureates is impressive, especially if considering the relatively small number of Soviet/Russian Nobel laureates.

Statues of Petr N. Lebedev (*left*, Chapter 1, A. K. Glebov) and Aleksandr G. Stoletov (*right*, S. I. Selikhanov) in front of the Faculty of Physics.

Aleksandr G. Stoletov (1839–1896) graduated from Moscow University in 1860 and stayed on to further his career in physics. He studied in Germany and France between 1862 and 1866. His research was in what we would call today condensed-state physics and materials properties. Upon returning to Russia he taught at Moscow University and received his professorial appointment in 1872. He organized weekly gatherings to discuss the latest developments in physics. He spent his summers in Western Europe to keep up with recent developments in his science. He founded a physical laboratory at Moscow University, which was the first of its kind in Russia. He was gifted in stimulating his students' interest in physics and for some time he was in charge of the physical section of the Polytechnical Museum.

Memorial plaque of Nikolai N. Bogolyubov on the façade of the Faculty of Physics.

The theoretical physicist Nikolai N. Bogolyubov (1909–1992) was a child prodigy. Already at the age of 19 he earned the PhD-equivalent Candidate of Sciences degree; and at 21 the DSc degree in mathematics. He worked at Kiev University between 1934 and 1959, and already from 1950, also, at the Steklov Institute and at Moscow University. Between 1950 and 1953 he was in charge of the mathematical section of Arzamas-16. He was a leading scientist at the United Institute of Nuclear Research in Dubna and was its director from 1965 until 1988.

Busts of Nikolai A. Umov (*left*) and Sergei I. Vavilov (*right*, Chapter 1) in the entrance lobby of the Faculty of Physics.

Nikolai A. Umov (1846–1915) majored in mathematics at Moscow University. After further studies, mainly in technology, he began his teaching career in 1871 at Novorossiisky University. In 1893, he was appointed professor of physics at Moscow University where after Aleksandr Stoletov's death, he took over Stoletov's course. Together with Petr Lebedev, Umov was active in organizing the Physical Institute at Moscow University. His research focused on condensed-state physics. For a long time he served as president of the Moscow Society of Naturalists.

Rem V. Khokhlov's bust (courtesy of Ekaterina Altova) at the Center for Non-linear Optics.

Rem V. Khokhlov (1926–1977) was a pioneer in non-linear optics. He graduated from Moscow University in 1948 and had a lightning career culminating in his appointment as Rector of Moscow University from 1974. He was a dedicated alpinist and died following an accident in the Pamir Mountains.

The Shternberg Astronomical Institute of Moscow University and the busts of Fedor A. Bredikhin (*left*) and Pavel K. Shternberg (*right*) in its entrance lobby; courtesy of Olga Dorofeeva.

The Sternberg Astronomical Institute at 13 Universitetsky Avenue is part of Moscow University. It operates in close cooperation with the astronomy section of the Faculty of Physics. Fedor A. Bredikhin (1831–1904) was an astronomer and professor of Moscow University. He was also the director of the Pulkovo Observatory. Pavel K. Shternberg (1865–1920) was also an astronomer, but more importantly for the Institute bearing his name, he was a revolutionary politician and participant in the Civil War.

The seeds of one of today's most prestigious universities of Russia were planted at the Faculty of Physics. It is now the Moscow Institute of Physics and Technology (*PhysTech*). Leading scientists already in 1938 recognized the need of training professionals for solving special problems of national interest. The idea was to create an independent institution of higher education involving research institutes of the Academy of Sciences. This could become a reality only after WWII when the need for such an institution had become yet more pressing due to the atomic bomb project. Top physics professors became members of the initial governing body, such as Abram Alikhanov, Sergei Vavilov, Ivan Vinogradov, Petr Kapitsa, Igor Kurchatov, Nikolai Semenov, and Sergei Khristianovich. Kapitsa had a vision of a school outside the capital city though not too far away, something like a Soviet Cambridge or Oxford. Also, Kapitsa followed Abram Ioffe's (Chapter 1) organization in Leningrad in which top students were being trained jointly at the Polytechnic Institute and the Institute of Physical Technology of the Academy of Sciences. In the Moscow version, initially, this training was accomplished in the framework of a new Faculty of Physics and Technology of Moscow University.

Eventually, when the institution became independent, its location was established outside of Moscow, in Dolgoprudny, Moscow Region.[4]

Faculty of Chemistry

Central section of the Faculty of Chemistry.

Statues of Dmitry I. Mendeleev (*left*, Chapter 1) and Aleksandr M. Butlerov (*right*) in front of the Faculty of Chemistry.

Aleksandr M. Butlerov (1828–1886) spent his scientific career in Kazan, but was fully integrated in the international community of the leading chemists of his time. He was concerned with reactions and the syntheses of new organic compounds. He knew that the structure of chemical substances — the atomic connectivity and the spatial arrangement of atoms in a molecule — determines their properties and behavior. In the Soviet Union, in an exaggerated way, he

[4] Another top institution of higher education, the Moscow State Institute of International Relations, had its origin as a component of Moscow University in the second half of the 1940s.

was considered as the creator of structure theory. The Soviet ideologues used his teachings to attack milestone theoretical advances by scientists in the West, such as Linus Pauling and his theory of resonance.

The theory of resonance used models of extreme structures whose fast inter-conversion was supposed to represent the real structure. The Soviet ideologues considered non-existing structures even for modeling to be alien to Marxism-Leninism. Their criticism had fast become political with tragic consequences that included the humiliation of internationally renowned scientists and some of them losing their jobs. The background of the attacks was Stalin's ruthless anti-science campaign and the Soviet Union was becoming increasingly isolated from the rest of the world.

One of the most prominent scientists who lost his job in this affair was Vladislav V. Voevodsky (1917–1967). He was a physicist engineer and one of Nikolai Semenov's favorite pupils. Voevodsky's main interest was in studying the relationship of structure and the mechanism of reactions and in utilizing physical techniques in chemistry. He worked in Semenov's Institute and at Moscow University, but in 1952, he was fired from the University for using Pauling's theory of resonance. From 1953 until 1961, he taught at the Moscow Institute of Physics and Technology, which was apparently more concerned with the quality of teaching than with ideological purity. In 1961, Voevodsky moved to Novosibirsk and was one of the top professors in the new science center. He received the State Prize in 1968 (posthumously).

There are three memorial tablets (not shown) on the façade of the Faculty of Chemistry at the entrance. They are for Aleksandr Nesmeyanov, Nikolai Semenov, and Nikolai Emanuel. All three worked at the Faculty of Chemistry — parallel to other places of employment — from the time of its opening in 1953 throughout their careers.

Busts of Vladimir V. Markovnikov (*left*) and Vladimir I. Vernadsky (*right*, Chapter 3) in the entrance lobby of the Faculty of Chemistry.

Vladimir V. Markovnikov (1838–1904) was an organic chemist, especially interested in petroleum chemistry. He synthesized new substances and investigated the mechanism of their reactions. Some of these reactions have been named after Markovnikov. He graduated from Kazan University and continued his career there, eventually as a professor. He studied at some of the best chemical laboratories in Germany. In 1871, he left Kazan University in protest against the firing of one of his colleagues. He moved first to Novorossiisky University, then to Moscow University where he was in charge of the chemical laboratory. Eventually, Nikolai Zelinsky followed him in this position. As Markovnikov attended Butlerov's courses in Kazan and Zelinsky continued his work in Moscow, this was a nice succession of some of the most outstanding Russian chemists.

Memorial of Klavdiya V. Topchieva (1911–1984). She was a student, a professor and ultimately the dean of the Faculty of Chemistry. Her main interest in research was in catalysis.

The Faculty of Chemistry was a cradle of polymer science and technology in the Soviet Union. When Valentin A. Kargin (1907–1969), started his career, it was not yet a generally accepted view that macromolecules — another name for polymers — existed at all. Many believed that what others thought were macromolecules were in reality systems of colloids — mixtures of microscopically dispersed particles of one substance suspended in the medium of another substance. By the mid-1950s, the existence of polymers had become firmly accepted in the world literature. Kargin started his association with Moscow University at the department of colloid chemistry and eventually founded a department of polymers. This was at the time when there was no polymer science let alone polymer industry yet in the Soviet Union.

Private initiatives were not tolerated at the time, but Kargin succeeded in convincing a great authority of science, Nikolai Semenov, that there should be polymer science. Semenov managed to tell the supreme leader Nikita Khrushchev about the importance of polymers. Khrushchev asked Semenov to prepare a report for the leadership of the Communist Party. This is how polymer science and industry was begun in the Soviet Union.

Kargin's department of polymer science received a separate building in 1966; alas, Kargin did not live long, and worked only a few years there. One of his pupils became his successor, Viktor A. Kabanov (1934–2006), who headed the department for 37 years. Kabanov has his own memorial tablet next to Kargin's. There is another tablet of Nikolai A. Plate (1934–2007), a chemistry graduate of Moscow University who served also as director of the Topchiev Research Institute of Petrochemical Synthesis. A joint tablet remembers Mikhail A. Prokofiev (1910–1999) and Zoya A. Shabarova (1925–1999), both natural products chemists, who were among the pioneers of nucleic acid research in the Soviet Union. Prokofiev held appointments of deputy minister and minister of education of the Soviet Union between 1951 and 1984.

The original building of the polymer department has been expanded and it now houses, in addition, the Belozersky Research Institute of Physical-Chemical Biology of the University. The scientific interests of the two organizations overlap considerably.

Memorial plaques of Andrei N. Belozersky (*left*) and Izrail M. Gelfand (*right*).

Andrei N. Belozersky (1905–1972) was orphaned early, but he was determined to study and acquire a good education. He graduated in physiology from the University in Uzbekistan in Tashkent. In 1930 he joined the new department of plant physiology at Moscow University. He researched substances formed jointly by nucleic acids and proteins. In 1960, he was appointed head of the department of plant biochemistry at the Faculty of

Biology. Eventually, he created the inter-faculty Laboratory of Bioorganic Chemistry, which then became the Belozersky Institute.

Izrail M. Gelfand (1913–2009) was one of the great mathematicians of the 20th century. He had a professorial appointment in what is now the Belozersky Institute between 1965 and 1991, where he organized a famous biological seminar. He had other jobs simultaneously and taught mathematics at Moscow University since 1935. He used to have an appointment at the Steklov Institute, but he left it for the Keldysh Institute of Applied Mathematics.

Like most scientists in the Soviet era, he was not allowed foreign travel and he was one of the most conspicuous world-famous scientists treated in such a way. There was also an anti-Semitic side of this story, because various restrictions were especially rigorously exercised towards Jewish scientists. By the time Gelfand was elected a full member of the Soviet Academy in 1984, he had already become foreign member of the US National Academy of Sciences (1970), the Royal Swedish Academy of Sciences (1974), the French Academy of Sciences (1976), and the Royal Society (London, 1977), and he had received the first Wolf Prize in Mathematics (1978). When he could finally travel in 1989, the 76-year-old Gelfand moved to the United States and built up a strong research program at Rutgers University in New Jersey.

Faculty of Biology

Central section of the Faculty of Biology, Faculty of Soil Science, and other institutions. Across the street, there is the Botanical Garden of the University.

We introduce the Faculty of Biology through a few of its outstanding representatives, starting with eight that have their portraits in the entrance lobby (only three of the eight are reproduced here). One of the eight, Ivan M. Sechenov, is introduced in Chapter 5.

Portraits, *from left to right*: Aleksei N. Severtsov, Mikhail A. Menzbir, and Nikolai M. Kulagin.

Aleksei N. Severtsov (1866–1936) graduated from Moscow University in 1889 specializing in zoology. From 1893 he taught at his Alma Mater. He did postgraduate studies in Western Europe from 1895 until 1898. Upon his return to Russia he won a professorial appointment in what is now Tartu, Estonia (at the time its name was Yuryev). Then, he continued at Kiev University. He founded evolutionary morphology in Russia.

Mikhail A. Menzbir (1855–1935) graduated from Moscow University in 1878 and continued his career there. He completed two years of studies in Western Europe and in 1884 he started giving lectures in the department of comparative anatomy at Moscow University. He was rising on the university ladder, but in 1911 he was among the professors who left the university in protest against the violation of university autonomy by the authorities. He continued teaching at the Higher Courses for Women. He returned to Moscow University in 1917. He founded and led a laboratory of zoogeography and comparative anatomy both at Moscow University and at the Science Academy. He organized and directed an institute and museum of comparative anatomy for the Academy of Sciences. He chaired the Moscow Society of Naturalists for twenty years, to the end of his life.

Dmitry N. Anuchin (1843–1923) graduated from Moscow University in 1867 and became a multifaceted scientist and pedagogue. His fields included geography, anthropology, ethnography, and archeology. He initiated what is now the Anuchin Museum of Anthropology at the University as well as other museums. He became professor and head of the department of geography and ethnography, and mentored future leading scientists. There is a memorial plaque on the façade of his former home, between 1911 and 1923, at 9 Khlebny Street in downtown Moscow.

Petr P. Sushkin (1868–1928) zoologist graduated from Moscow University in 1889 and started his career at the department of comparative anatomy at his Alma Mater. His main interest was in ornithology. After having earned his higher degrees, he received professorial appointments and worked at a succession of universities in Kharkov, Simferopol, and Leningrad and at the Zoological Museum of the Academy of Sciences.

Nikolai Yu. Zograf (1851–1919) started his career at the Zoological Museum following his graduation from Moscow University. He taught at a variety of schools, including the Novorossiisky University. Finally, he was appointed professor of zoology at Moscow University in 1888. He participated in expeditions, chaired the section of ichthyology of the Society for the Acclimatization of Animals and Plants, and fulfilled other societal functions in science.

Nikolai M. Kulagin (1860–1940) was a zoologist and apiculturist. Upon graduation from Moscow University, he started at the Zoological Museum. He then received a professorial appointment at Moscow University, was director of the Moscow Zoopark, and taught at the Moscow Agricultural Institute. Along with many others, he resigned from his position at Moscow University in 1911 in protest. In 1919, Kulagin returned to Moscow University and from 1925 he worked as head of the department of entomology until the end of his life. He was dedicated to nature preservation and was active in the movement from 1906. Between 1918 and 1931 he was in charge of applied zoology at the Polytechnical Museum.

Nikolai V. Nasonov (1855–1939) graduated from Moscow University in 1879 and his first job was at the Zoological Museum. He had a professorial appointment at Warsaw University. He was director of the Zoological Museum of the Academy of Sciences between 1906 and 1921; of the Laboratory of Experimental Zoology between 1921 and 1931; and of a laboratory at the Institute of Cytology, Histology, and Embryology from 1931 until the end of his life.

There are a number of memorial tablets at the Faculty of Biology. We mention four erected in the entrance lobby. Andrei N. Belozersky (see above) is remembered as a founder of molecular biology in Russia. Mikhail V. Gusev (1934–2005) graduated from Moscow University in 1957. His research focused on cell physiology and the mechanism of photosynthesis in microorganisms. He was dean of the Faculty of Biology for 33 years (1973–2005), which included the most turbulent times in politics, but his position proved unshakeable. Sergei E. Severin (1901–1993) was a biochemist who was active in directing various research institutes of the Academy of Medical Sciences. He held leading positions in biochemical organizations. For more than fifty years he was the founding head of the department of animal biochemistry at the Faculty of Biology. Vladimir E. Sokolov (1928–1998) graduated from Moscow University in 1950 and worked at his Alma Mater from 1956. From 1967, he was also director of what is now the Severtsov Institute of Ecology and Evolution (33 Leninsky Avenue). His research was in zoology and he popularized biology.

Bust of Vladimir Vernadsky (E. M. Vilensky) at the Earth Science Museum of Moscow University.

3

Earth Science Museum

The rector, Aleksandr Nesmeyanov, initiated the creation of the Earth Science Museum around 1950, simultaneously with planning the move to the new campus on Vorobyovy Hills,[1] A large collection came together from individuals, organizations, and institutions. It has continuously expanded from expeditions of associates of the University. A special fund was established to commission busts of scientists, naturalists, and explorers from the best sculptors of the country.

The research work by the associates of the Museum and the educational activities by all participating instructors of the University complement each other. The Museum functions as a laboratory for classes as part of a diverse curriculum of subjects, including geography, geology, anthropology, biology, environmental science, soil science, and others. The Museum, which is a museum of natural history, can be visited in organized groups by prearrangement.

The Earth Science Museum is on floors 24–28, in the top rectangular portion of the tower.

[1] The area of Vorobyovy Hills was called Lenin Hills in Soviet times; today, it is Vorobyovy Hills again. The official address of the University has remained 1 Lenin Hills.

Rotunda

Part of the Rotunda as viewed from the 32nd floor.

The top part of the university tower is the Rotunda. It has a festive hall for elegant celebrations with a seating capacity of about 50 people. There are a number of busts, other artifacts, and displays, narrating the history of the University. A few examples are shown here.

Left: Empress Elizabeth (in Russian Elizaveta Petrovna (1709–1762; M. Galliulin, 2012), the daughter of Petr I and Ekaterina I, was Russian Empress from 1741. She decreed the establishment of Moscow University in 1755. *Middle*: Ivan I. Shuvalov (1727–1797; copy, possibly by Santino P. Kampioni) assisted founding Moscow University. *Right*: Karl F. Rulie (1814–1858) was a biology professor. His interest was in geology, evolutionary paleontology and in the influence of environment on animal development.

An unusual image of Mikhail Lomonosov (I. Kotov, 1971).

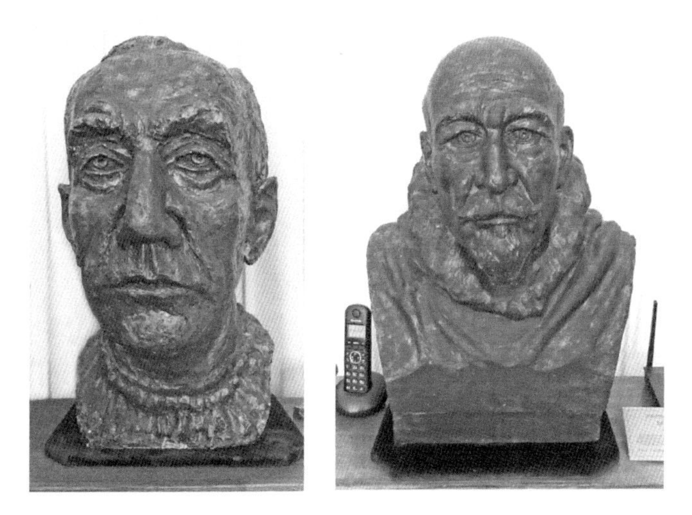

Left: Roald Amundsen (1872–1928; I. A. Velmina, 1997) Norwegian polar explorer. *Right*: Nikifor A. Begichev (1874–1927; B. N. Brodsky, 1962) polar explorer.

Views from the 32nd Floor

Views from the 32nd floor. *Left*: The Faculty of Physics (lower right) and the building behind it, to the left, includes a division of computational technology. *Right*: A new section of town, the so-called Moscow City.

The Museum Floor by Floor

The Museum itself occupies Floors 24–28 of the Main Building consisting of 24 halls and occupying 25 hundred square meters (27 thousand square feet). Each floor has a general topic and our narrative follows the floors in descending order. The subdivision of topics among the halls of each floor is not rigorous. A special exposition "Our Earth in the Universe" complements them on Floor 30.

Floor 28: Geo-dynamics and Endogenous Processes

Two images of Floor 28; one displaying a globe in Hall 5; the other, Charles Darwin's statue (S. T. Konenkov) in Hall 7.

Hall 3 Structure and Evolution of the Earth

Peter-Simon Pallas (L. M. Pisarevsky).

Peter-Simon Pallas (1741–1811) was a German-born Russian polymath. He studied in Berlin, Halle, and Gottingen; and learned botany according to Karl Linné's taxonomy. Pallas earned his doctorate in medicine from Leiden University. In 1767, he and his family moved to St. Petersburg where he had a professorial appointment. He organized expeditions in Russia and collected information in a broad range of natural sciences. He was an encyclopedist and published papers in zoology, botany, paleontology, mineralogy, geology, and other areas.

Hall 4 Magmatic Formations

Alexander von Humboldt's bust at the Museum (*left*, D. I. Derunov) and his statue at Humboldt University in Berlin (*right*, Reinhold Begas, 1883).

Alexander von Humboldt (1769–1859) was a German naturalist, explorer, and the founder of modern geography. His many memorials in a number of countries attest to his international fame.

From left to right: Aleksandr Karpinsky (P. V. Kenig, Chapter 1); Stepan Krasheninnikov (A. L. Stepanyan); and Frants Levinson-Lessing (V. V. Miklashevskaya).

Stepan P. Krasheninnikov (1711–1755), explorer, botanist, and ethnographer, researched Siberia and Kamchatka. Vladimir Vernadsky compared Krasheninnikov to Lomonosov in that their activities completed the preparatory period of Russian scientific creativity and made Russia an equal party among educated mankind.

Frants Yu. Levinson-Lessing (1861–1939) geologist and petrographer studied and worked mostly in St. Petersburg/Petrograd/Leningrad. A "charming" story: In the 1920s, as Levison-Lessing's text on petrography was undergoing checking, an overzealous censor replaced the genuine Russian word "petrografiya" by "leningrafiya."[2]

Hall 5 Geo-tectonics

Boris Golitsin (N. B. Nikogosyan) and Vladimir Obruchev (D. P. Shvarts).

[2] https://ru.wikipedia.org/wiki/Левинсон-Лессинг,_Франц_Юльевич; 10/25/16.

Boris B. Golitsin (1862–1916) was a naval officer turned physicist and inventor. He contributed to the development of geophysics, invented an electromagnetic seismograph, and helped establish the seismographic service in Russia. He was elected foreign member of the Royal Society (London).

Vladimir A. Obruchev (1863–1956) studied at the Mining Institute in St. Petersburg and participated in expeditions to the Ural Mountains and Central Asia. He taught at the Tomsk Polytechnic Institute and elsewhere before joining the Moscow Mining Academy. He moved to Leningrad, assumed important functions in the Academy of Sciences and in the Geographical Society, and wrote works popularizing science.

Left: Eduard Suess (B. D. Korolev).[3] *Right*: Suess's bust (1928) in Schwarzenberg Square in Vienna.

Eduard Suess (1831–1914), Austrian geologist, researched the geography of the Alps. He was professor of paleontology, and later, of geology at the University of Vienna. He was a pioneer in taking into account ecological considerations in scientific discussions.

Mikhail Usov (D. S. Shaposhnikov).

[3] A copy of this bust stands in Hall 13.

Mikhail A. Usov (1883–1939) was Obruchev's student. Usov's research interest was in the evolution of Earth. He studied at the Tomsk Polytechnic Institute and participated in expeditions. Eventually, he returned to Tomsk and became a leading professor at his Alma Mater. He was active in the application of earth sciences in the development of the Siberian economy.

Hall 7 Methods of Geological Research

Charles Darwin (S. T. Konenkov).

Charles Darwin (1809–1882), the English naturalist and world-famous scientist of evolution, has been a much revered icon in Russia. There is a State Darwin Museum in Moscow — a natural history museum (Chapter 8).

Aleksei Pavlov (Evgeniya M. Kovarskaya).

Aleksei P. Pavlov (1854–1929) studied at Moscow University, became a geologist and paleontologist, and taught at his Alma Mater and elsewhere. His geological studies focused on the surroundings of Moscow. He and his geologist wife, Mariya V. Pavlova (Chapter 8), organized the Geological Museum of Moscow University.

Floor 27: Formation of Minerals

Classroom exercise in Hall 8.

Left: Busts of Aleksandr Fersman and Vladimir Vernadsky in Hall 9. *Right*: Samples of minerals on display in Hall 12.

Hall 8 Processes of Mineral Formation

Left: Abu-Reikhan Al-Biruni (N. P. Khodamaev). *Right*: Al-Biruni statue in Tehran (courtesy of Behrooz and Mehdi Esrafili).

Abu-Reikhan Al-Biruni (973–1048) was a polymath of the golden era of Islam. He was an Iranian originating from what used to be Soviet Central Asia. He lived also in the region of present-day Afghanistan. He was not only a scientist, but also a traveler, explorer, and writer. His fields of inquiry included, among many others, the earth sciences, and sometimes he is referred to as "the father of geodesy." He explained the different phases of the moon, and his statue in Tehran refers to his discoveries in astronomy.

Nikolai V. Belov (1891–1982; N. V. Bogushevskaya) was a crystallographer at the Institute of Crystallography and a professor at Moscow University.

Evgraf Fedorov (A. V. Grubbe) and Vasily Severgin (S. D. Popov).

Evgraf F. Fedorov (1853–1919) was a crystallographer and mathematician and director of the Mining Institute in St. Petersburg. His best-known achievement was the rigorous derivation of the 230 three-dimensional space groups of symmetry. As it often happens in science, independent of Fedorov, the English William Barlow and the German Arthur M. Schoenflies also derived these space groups. Fedorov was voted for and declined full membership in the Academy of Sciences. He was elected an "adjunct" member in 1901, but in 1905 he resigned from it accusing the Academy of hindering progress.

Vasily M. Severgin (1765–1826) continued Lomonosov's cause and advanced chemical and geological knowledge in Russia. He was talented in the arts, including music, but acquired his principal education in chemistry and mineralogy. He repeated Antoine Lavoisier's experiments and popularized them among Russian chemists. He introduced Russian nomenclature for many chemical terms and created chemical dictionaries in Russian. He conducted expeditions and was one of the initiators of the Mineralogical Society in St. Petersburg.

Hall 9 Processes of Ore Formation

Vladimir Vernadsky (E. M. Vilensky).

Vladimir I. Vernadsky (1863–1945), scientist, geologist, and thinker, created a strong school of geosciences. He put great emphasis on chemistry in the applications of the achievements of earth sciences for the national economy. He recognized the importance of Space in forming his world views.

Vernadsky earned his doctorate in 1897 at St. Petersburg University and Mendeleev was one of his professors. In 1908, Vernadsky visited France and Great Britain. In 1922, he participated in creating the Radium Institute and directed it through 1939; today it is the Khlopin Radium Institute in St. Petersburg. In the period 1922–1926, Vernadsky was in Paris to give courses and conduct research at the Sorbonne. In 1921, he was briefly arrested on charges of espionage but soon freed as a result of protests by leading scientists. On the other hand, even during the darkest years of Stalin's terror in the 1930s, Vernadsky was left alone, just as Fersman and Karpinsky were.

Vernadsky actively participated in modernizing the Soviet economy. He supported the plans to electrify the whole country (the so-called GOELRO Plan). In 1940, he demonstrated exceptional foresight in initiating expeditions for discovering locations for mining uranium. He understood the potential of the forces of the atomic nucleus for making bombs and producing energy. There is a Vernadsky Geological Museum at 11 Mokhovaya Street (Chapter 8).

Aleksandr Fersman (A. V. Babichev).

Aleksandr E. Fersman (1883–1945) was a geochemist and a founder of geochemistry whom the writer Aleksey Tolstoy called "the poet of stones."[4]

[4] https://ru.wikipedia.org/wiki/Ферсман,_Александр_Евгеньевич; 10/25/16.

Fersman completed his studies at Moscow University as one of Vernadsky's pupils. As professor of geochemistry at Moscow University, Fersman gave the first ever course of geochemistry. He conducted exploratory expeditions to many regions of the Soviet Union looking for natural resources to help the national economy. From 1917 until his death, he was the director of what is today the Fersman Mineralogical Museum, 18 Leninsky Avenue (Chapter 8).

Left: Sergei S. Smirnov (1895–1947; M. K. Anikushin), a Leningrad geologist and specialist in locating mineral ores. *Right*: Aleksei I. Tugarinov (1917–1977; V. V. Miklashevskaya) geochemist specialized in the chemistry of rare and radioactive elements.

Hall 10 Occurrence of Metals

From left to right: Grigory Shchurovsky (G. M. Toidze); Vasily Tatishchev (M. E. Yaroslavskaya); and his equestrian statue (A. I. Rukavishnikov, 1998) on the bank of the Volga River in Tolyatti (Togliatti).[5]

[5] https://ru.wikipedia.org/wiki/Файл:Monument_of_Tatishchev_in_Togliatti.jpg, public domain; 10/25/16.

Grigory E. Shchurovsky (1803–1884) was a geologist, paleontologist, and anatomist. He completed his studies in the medical school of Moscow University, became the first professor of geology and mineralogy at his Alma Mater, and stayed in charge of his department for half a century.

Vasily N. Tatishchev (1686–1750) authored the first comprehensive treatise about the history of Russia and was also a geographer and economist. As a statesman, he contributed to the annexation of the Ural Mountains and to the founding of important cities, such as Ekaterinburg (it was Sverdlovsk between 1924 and 1991) and Perm. Tatishchev's equestrian monument recognizes his activities as a statesman.

Hall 11 Occurrence of Non-metallic Ores

Nikolai I. Koksharov (1818–1893; M. A. Gritsuk) described the crystal morphology of many minerals.

Hall 12 Occurrence of Fuel Ores

Ivan M. Gubkin's (1871–1939; G. V. Neroda) career as geologist is described in Chapter 4.

Floor 26 Exogenous Processes and History of Earth

Displays in the Halls 13, 14, and 15. *From the upper left, clockwise*: consequences of weathering; corals; tyrannosaurus; and ammonite.

Hall 13 Exogenous Processes

From left to right: Dmitry N. Anuchin (L. N. Lavrova, Chapter 2). Aleksandr A. Kruber (1871–1941; R. G. Geondzhian), a geographer at Moscow University with main interest in physical geography and cartography. Evgeny V. Milanovsky (1923–2012; A. A. Manuilov) geology professor at Moscow University researched tectonics and volcanology.

Oktavy Lange and Mikhail Sumgin (both by N. A. Velmina).

Oktavy K. Lange (1883–1975) originated from a German-Italian family, was a geologist, and the founder of engineering geology. He studied at Moscow University and was later head of the hydrogeology department at the Mining Academy before moving on to Central Asia. During the last three decades of his life he was back at his Alma Mater, first in charge of the department of dynamic geology, then, of hydrogeology.

Mikhail I. Sumgin (1873–1942) was one of the founders of the science of permafrost. He was in a commission, initiated by Vernadsky, and then in a research institute, named after V. A. Obruchev, for studying rock formations of minerals under the conditions of permafrost or seasonal frost.

Hall 14 Sea Activities

Nikolai Knipovich (A. A. Stempkovsky) and Fedor Litke (L. E. Kerbel).

Nikolai M. Knipovich (1862–1939) was a biologist with main interest in ichthyology, zoology, and hydrobiology. He participated in expeditions to the White Sea and the Murmansk area to assess their value for the economy and to investigate their flora and fauna.

Fedor P. Litke (1797–1882) was orphaned and began a career in the Russian Navy through family connections. His sea voyages fostered his interest in organizing scientific expeditions. He was among the leaders of the Russian Geographical Society and served also as President of the Academy of Sciences (1864–1882).

Stepan Makarov (V. S. Chebotarev) and Yuly Shokalsky (S. I. Fokin).

Stepan O. Makarov (1849–1904) admiral and oceanographer researched the Pacific Ocean and the Arctic, conducted expeditions, including those around the globe, and initiated the development of icebreakers. He was killed in a battle during the Russo-Japanese war.

Yuly M. Shokalsky (1856–1940) became an orphan at an early age. He studied at the Naval Academy and had a career in geography and in teaching oceanography and related subjects. He was active in the Russian Geographical Society. He assisted *Gosplan* — the ministerial institution for working out the five-year plans for the Soviet economy.

Hall 15 Ancient History of Earth

From left to right: Aleksei A. Borisyak (A. Allakhverdyants). Fedosy N. Chernyshev (1856–1914; Z. N. Rakitina) geologist and paleontologist researched the stratigraphy of the Ural Mountain. Aleksandr N. Mazarovich (1886–1950; A. S. Rabin) hydro-geologist studied and taught at Moscow University and the Mining Academy.

Aleksei A. Borisyak (1872–1944) was a graduate of the Mining Institute in St. Petersburg. Eventually, he taught at the same institution (by then, it was Leningrad). He then founded the department of paleontology at Moscow University. He was also the founding director of what is today the Borisyak Institute of Paleontology (Chapter 8). He contributed to the theory of the tectonic structure of the Earth's crust.

Vladimir Kovalevsky (T. N. Ozolina). The postage stamp depicts his wife, the mathematician Sofiya Kovalevskaya.

Vladimir O. Kovalevsky (1842–1883) was a biologist and paleontologist. The future renowned mathematician Sofiya V. Kovalevskaya entered a marriage of convenience with him so that she could go abroad to attend college. Eventually, their union became a true marriage. Kovalevsky's research supported Darwin's teachings about evolution. His brother, Aleksandr, was an embryologist.

Hall 16 Appearance of Humans on the Earth

Left: Nikolai I. Andrusov (1861–1924; A. I. Teneta) geologist and paleontologist worked in stratigraphy, paleontology, and oceanography. *Right*: Sergei N. Nikitin (1851–1909; P. I. Bondarenko) geologist organized the geological survey of Russia.

Petr Kropotkin (N. I. Niss-Goldman) and Georgy Mirchink (M. F. Listopad).

Petr A. Kropotkin (1842–1921) originated from an aristocratic family and was a revolutionary-anarchist as well as a geographer and historian. For decades he lived in exile in Western Europe and returned to Russia in 1917. His main scientific interest was in physical geography. His scientific activities focused on Siberia, and he coined the expression "eternal frost." One of the busy Moscow subway stations bears his name, "Kropotkinskaya," for his revolutionary activities.

Georgy F. Mirchink (1889–1942) was a geologist whose main interest was in quaternary geology, i.e., in the geology of the latest period in the history of Earth preceding the current age. He studied and then worked at Moscow University and won international recognition for his science. In 1941, he was arrested on false charges and died in prison.

Floor 25: Natural Zones

Left: Hall 19 with the bust of Konstantin Gedroits and a display of the fauna of semi-deserts. *Right*: Elk in Hall 18. Both photographs courtesy of Konstantin Skripko.

Hall 17 (and part of Hall 20) Natural Zones and Their Components

Vasily Dokuchaev (I. M. Chaikov).

Vasily V. Dokuchaev (1846–1903) was a geologist turned soil scientist and he initiated the investigation of the nature and origin of soil. He was also an explorer and conducted expeditions. There is another Dokuchaev bust in the Promenade of Scientists (Chapter 2).

From left to right: Vladimir L. Komarov (N. K. Ventsel, Chapter 1). Pavel A. Kostychev (1845–1895; I. M. Chaikov) agronomist studied the role of microorganisms in humus formation. Dmitry N. Pryanishnikov (O. V. Kvinikhidze, Chapter 6).

Left: Nikolai M. Sibirtsev (1860–1900; S. V. Kazakov) geologist and soil scientist researched the genetic relationships of soils. *Right*: Kliment A. Timiryazev (Z. V. Bazhenova, Chapter 6).

Hall 18 From Tundra to Woods

Vasily Alekhin (I. L. Yakovleva) and Semen Dezhnev (B. N. Brodsky).

Vasily V. Alekhin (1882–1946) studied and was professor at Moscow University. His field was botany and plant physiology. He founded and directed a new chair of geo-botany at the Faculty of Biology. He also taught at other schools in Moscow, among them, the Mining Academy, the Higher Courses for Women, and the Timiryazev Academy.

Semen I. Dezhnev (1605–1673) was an explorer of Northern and Eastern Siberia. As recognition of his explorations, the easternmost point of the Eurasian continent, on the Chukotka (Chukchi) Peninsula, is called Dezhnev Cape.

From left to right: Lev S. Berg (1876–1950; E. S. Gerlenshtein) researched ichthyology, geography, and the theory of evolution. Georgy F. Morozov (1867–1920; L. Ya. Doronina) had an interest in forestry. Nikolai A. Severtsov (I. A. Rabinovich, Chapter 2).

Hall 19 Wooded Steppes; Steppe Semi-deserts

From left to right: Vasily R. Vilyams [Williams] (A. I. Sergeev, Chapter 6). Konstantin Gedroits (A. D. Korsukov). Ivan V. Michurin (S. D. Lebedeva, Chapter 1).

Konstantin K. Gedroits (1872–1932) studied the chemistry of the soil. He improved the methods of chemical analysis and opened new ways of enhancing soil quality with chemistry. In 1932, he was arrested and he died in the same year.

Hall 20 Deserts, Subtropics, Hot Climate, High Altitudes

Left: Sergei A. Zakharov (1878–1949; A. P. Vishkarev) soil scientist studied at Moscow University and participated in expeditions. *Right:* Sergei S. Neustruev (1874–1928; L. M. Kapinus) organic chemist became an explorer and soil scientist.

Floor 24: Physical-geographical Areas

Hall 24 with Andrei Krasnov's bust.

Hall 21 Russian Plain, Ural, Crimea, Carpathian Region

Left: Aleksandr A. Borzov (1874–1939; P. V. Kenig) geographer helped organize a research institute and museum of geography at Moscow University. *Right*: Ippolit M. Krasheninnikov (1884–1947; M. B. Aizenshtadt) botanist worked at the Botanical Gardens of St. Petersburg and Moscow.

Petr Rychkov (I. A. Rabinovich) and Aleksey Tillo (P. I. Bondarenko).

Petr I. Rychkov (1712–1777) was a civil servant with interest in geography and wrote about the Southern Ural region. His son, Nikolai P. Rychkov (1746–1798), was an explorer and geographer who initiated breeding of silkworms in Russia.

Aleksei A. Tillo (1839–1899) geographer, cartographer, and geodesist, combined his service in the military with scientific activities. He determined the levels of the Caspian Sea and the Aral Sea and investigated terrestrial magnetism. He organized expeditions to the upper reaches of the major rivers in European Russia.

Hall 22 General Overview of Russia and the World. Caucasian Region, Central Asia

From Left to right: Herman von Abich (1806–1886; N. I. Kuznetsov), a German mineralogist, researched the geology of Russia. Dmitry N. Anuchin (A. V. Prisyazhnyuk, Chapter 2).[6] Aleksei P. Pavlov (1854–1929, T. R. Polyakova) geologist and paleontologist, professor of Moscow University and founder of the Moscow Geological School.[7]

[6] There is another Anuchin bust in Hall 13.

[7] There is another Pavlov bust in Hall 7.

Petr Semenov-Tyan-Shansky (M. L. Litovchenko) and Aleksandr Voeikov (Yu. G. Neroda).

Petr P. Semenov-Tyan-Shansky (1827–1914) geographer, botanist, and statistician acquired the added name "Tyan-Shan" after his visit to the Tian-Shan mountains where he was the first European explorer. He had a number of functions and positions in learned societies and in the broader society. He collected important items from nature and valuable works of art. Five of his children became noted professionals in geography, ethnography, meteorology, and related areas.

Aleksandr I. Voeikov (1842–1916) was a meteorologist, climatologist, and an active proponent of vegetarianism. He studied in St. Petersburg and continued his studies in Western Europe. He investigated various locations in Russia to form a comprehensive picture about Russia from the point of view of meteorology and for assisting agriculture.

Left: Aleksei P. Fedchenko (1844–1873; E. S. Yarosh) parasitologist, entomologist, and explorer of Central Asia and the Pamir Mountain lost his life in an expedition. *Right*: Nikolai M. Przhevalsky (1839–1888; A. V. Pekarev), an army officer, investigated the flora, fauna, and the climate in Northern Tibet and other regions.

Hall 23 Siberia and the Far East

Ivan Chersky (A. V. Pekarev) and Gleb Vereshchagin (V. S. Manashkin).

Ivan D. Chersky (Jan Stanisław Franciszek Czerski, 1845–1892) was a Polish geographer and explorer. In 1863, he participated in an uprising against tsarist repression and was exiled to Siberia. He was prevented from receiving higher education and studied on his own, with friendly mentors. He published scientific treatises in geology and osteology of animals of Siberia. Scientific societies recognized his research contributions. He died during an expedition.

Gleb Yu. Vereshchagin (1889–1944) was a leading Soviet geographer, limnologist, hydro-biologist, and science organizer. Upon graduation from the University of Warsaw, he started his career at the Zoological Museum in Petrograd (as it was then). During the last 14 years of his life, he directed the Baikal Station of Limnology.

Left: Vladimir K. Arseniev (1872–1930; N. A. Pisarev) lived in Vladivostok and wrote books about his explorations of the Far East. His home has been turned into a museum. *Right*: Aleksandr F. Middendorf (1815–1894; L. K. Klimenkova-Krauze) geographer of Northern and Eastern Siberia. He investigated species preserved in frost.

Hall 24 Royal Stags and Parts of the World

Andrei N. Krasnov (L. D. Muravin).

Andrei N. Krasnov (1862–1915) was a botanist, soil scientist and paleontologist, the first in Russia who earned a doctorate in geography and defended his dissertation in an open defense at Moscow University. He organized expeditions to Eastern Russia and other Asian countries. He was an artist as well as a scientist.

Left: Gennady I. Nevelskoi (1813–1876; S. L. Ostrovskaya) admiral, researched the Far East and determined that Sakhalin is an island. *Right*: Nikolai N. Zubov (1885–1960; N. B. Nikogosyan) admiral in the Navy, oceanographer, and polar explorer, founded departments of oceanography.

*****

Bust of Mikhail Lomonosov in the Museum on the background of one of his famous statements. Its loose translation into English reads:

Dare now to give the inspiration
to Russia for gaining confidence
for delivering her own Platos
and wise people as Newton was.

Shukhov Tower (detail) at 18 Shukhov Street (photographed from Shukhov Street).

4

Technology and Technologists

This is a vast area with countless memorials in Moscow. Here, we present memorials of the most famous Russian engineer, Vladimir G. Shukhov, a few other outstanding individuals, a few institutions of higher education and research in technology, including the best-known Bauman University, and a sampler of memorials of the great successes in space exploration.

Vladimir Shukhov

Statue of Vladimir G. Shukhov (S. A. Shcherbakov and his group, 2008) on Turgenevskaya Square. The stand of the sculpture reflects the style of the Shukhov Tower.

Side panels of the Shukhov statue depicting characteristic structures and tools of his activities.

Vladimir G. Shukhov (1853–1939) was born into a family of nobility in Graivoron, in Western Russia, not far from Kharkov. He attended high school in St. Petersburg and for college, what is today Bauman University in Moscow. Nikolai Zhukovsky was among his instructors. Shukhov began his activities as an inventor during his student years. Upon graduation, in 1876, he was a member of the Russian delegation visiting the World Fare in Philadelphia.

Shukhov led a prolific life of a true polymath with versatile activities. His many achievements include a number of firsts in Russia and on a world scale. He designed and directed the development of the first petroleum processing plants in Russia, which at the time included the Baku oil fields. He designed and directed the building of oil pipelines, tankers, numerous bridges, and the steel lattice roofs of famous buildings. One of them is the GUM department store.

The GUM Department Store on Red Square and its steel lattice roof structure.

There are a number of other buildings in this book that have elements in their design related to Shukhov, such as the No. 2 University Clinical Hospital and the Museum of the History of Medicine (Chapter 5), the main building of the Moscow State Pedagogical Institute (Chapter 5), and the Study Hall No. 6 at the Timiryazev Agricultural Academy (Chapter 6).

Shukhov designed what was the world's first hyperboloid tower, a water tower, in Polibino of the Lipetsky Region, some 370 kilometers (230 miles) south/south-east from Moscow. Such structures consist of a series of hyperboloid sections stacked upon each other and from which a conical overall shape forms. These lattice constructions minimize wind load, which is especially advantageous for tall buildings. Hundreds of Shukhov towers were built in the Soviet Union.

Left: Shukhov designed the world's first hyperboloid tower for the 1896 Industrial and Art Exhibition in Nizhny Novgorod. After the exhibition, the tower was moved to Polibino, Lipetsky Region, where it still stands. The photograph shows the Tower in its original location (http://rss.zanostroy.ru/u/bayashnya.jpg; public domain; accessed 12/30/16). *Middle*: Another Shukhov Tower — the tower of the Oktryabsky radio center on Marshal Tukhachevsky Street as viewed from Narodnoe Opolchenie Street (courtesy of Olga Dorofeeva). *Right*:View of the Shukhov Tower at 18 Shukhov Street (photographed from the corner of Academician Petrovsky and Shabolovka Streets).

The most famous is the Shabolovka Tower in Moscow, known also as the Shukhov Tower, built originally for radio transmission. Its height is about 150 meters. This hyperboloid structure was envisioned originally for 350 meters, but there was not enough metal to build it at the time of the Civil War. Even the reduced height required an amount of metal that was not trivial to procure. An accident disrupted the work and Shukhov was accused of negligence and sentenced to execution by shooting. It was a conditional sentence reprieved until the completion of the tower. In 1922, upon the successful completion of the project, the sentence was annulled. From 1939, this tower was also used for television transmission.

The Shukhov Tower no longer fulfills its original functions and corrosion has left its mark. From time to time, the question of its disassembling had been raised. However, it represents great value as a memorial of the cultural heritage of Russia and it can be made corrosion resistant. There is a 1:30 scale model of the Shukhov tower at the Science Museum in London. Great architects, such as Antoni Gaudi, Le Corbusier, and Oscar Niemeyer, also used hyperboloid structures in their creations.

Mendeleev University

Relief of Dmitry Mendeleev (*left*) over the entrance to the initial building (*middle*) of the Mendeleev University of Chemical Technology, 9 Miusskaya Square, and the memorial plaque of Sergei Kaftanov on its façade (*right*).

The predecessor of Mendeleev University was founded at the end of the 19th century at the initiative of what is today Bauman University. Famous former students and professors included Sergei Kaftanov, Isaak Kitaigorodsky, Valery Legasov, and Aleksandr Topchiev.

Sergei V. Kaftanov (1905–1978) graduated from this School and he has a memorial plaque on the facade at 9 Miusskaya Square. His interest was in fuel chemistry and he taught general chemistry. He had a career in administration; he was minister of higher education, followed by his position as first deputy minister of culture. He ordered the dismissal of many serious scientists and their replacement by Lysenko's followers at the time of the anti-science terror. In 1962, he returned to Mendeleev University and served as its rector until 1973.

Aleksandr V. Topchiev (1907–1962) was an organic chemist with a principal interest in petroleum chemistry. He held high positions in the leadership of the Academy of Sciences and directed research institutes, among them, what is today the Topchiev Institute of Petrochemical Synthesis, 29 Leninsky Avenue.

Mendeleev statue in the Miusskaya Square building (courtesy of Maya Sirotina).

It illustrates the international outlook of Mendeleev University that some international personalities have been awarded its honorary doctorate, Margaret Thatcher among them.

Detail of the memorial of Leonid A. Kostandov (Aleksandr Ryabichev, 2002) on the façade of Moscow State University of Environmental Engineering (courtesy of Olga Dorofeeva).

There is a somewhat related memorial of Leonid A. Kostandov (1915–1984) on the façade of the main building of Moscow State University of Environmental Engineering at 21 Staraya Basmannaya Street. Kostandov started as a machinist in agricultural industry and worked his way up to becoming the minister of chemical industry of the Soviet Union (1965–1980). During the last few years of his life (1980–1984), he was the deputy chairman of the Soviet Council of Ministers. He is credited with substantial contribution to the development of Soviet chemical industry. He was buried in the Kremlin Wall, and he was the penultimate person accorded such an honor.

Karpov Institute

The Karpov Institute of Physical Chemistry at 10 Vorontsovo Pole Street is a research establishment outside of the system of the Academy of Sciences. It was one of the first research organizations under the Soviet rule. Lev Ya. Karpov (1879–1921) was a chemical engineer and revolutionary and he was a principal figure in the fledgling Soviet chemical industry. The Karpov Institute has had a broad scope of research projects and leading scientists worked for it at one time or another, such as Aleksei N. Bakh (Chapter 1), Aleksandr N. Frumkin (Chapter 1), Valentin A. Kargin (Chapter 2), Yakov K. Syrkin (Chapter 1), and many others.

Memorial plaques of Valentin A. Kargin (*left*, Chapter 2) and Sergei S. Medvedev (*right*) on the façade of the Karpov Institute (courtesy of Sergei Lakeev).

Sergei S. Medvedev (1891–1970) graduated from the University of Heidelberg, Germany, in 1914 and from Moscow University in 1918. He was an associate of the Karpov Institute of Physical Chemistry and was a professor at the Lomonosov Institute of Fine Chemical Technology. He contributed to the theory of polymerization and his results helped to develop the polymer industry.

Igor V. Petryanov-Sokolov (1907–1996) graduated from Moscow University and started research already during his student years. He joined the Karpov Institute in 1929 and worked there for his entire professional career. From 1945, he participated in the nuclear project. He was an innovator and created new materials of broad applications. He popularized science and co-founded journals for the dissemination of knowledge, among them, *Chemistry and Life*.

Yakov M. Kolotyrkin (1910–1995) graduated from Moscow University and the Karpov Institute was his first and only place of employment; eventually, he became its director. His main research interest was in corrosion and the electrochemistry of metals. Beneath the memorial plaques, there are tablets on the façade with quotations from Mikhail V. Lomonosov and Dmitry I. Mendeleev. We chose one by Mendeleev to reproduce here in English translation: "Knowing how much at ease, free and happy life in science is, there is a desire to see more people taking part in it."

Gubkin University

In 1930, the Soviet government ordered the creation of six new institutions of higher education in place of the Stalin Moscow Mining Academy, which was dissolved.[1] One of them was the Institute of Oil, named after Ivan Gubkin and he was appointed to be its rector. Some of the founders of the Soviet oil and gas chemistry and chemical industry were among the professorial staff, such as Sergei Nametkin, Leonid Leibenzon, and Aleksandr Topchiev. Today, this is the Gubkin University of Oil and Gas — a national research university.

[1] Nowadays the opposite trend is observed in that huge institutions have been created by joining heretofore independent institutes of higher education, which are more or less related thematically and geographically, but sometimes may even be quite remote in both aspects.

The main building of Gubkin University of Oil and Gas at 65 Leninsky Avenue and decoration of the fence in the form of an oil well.

Left: Gubkin's statue (2011) in front of the main building of Gubkin University. *Right*: Relief on the façade of the main building depicts Gubkin and industrial apparatus.

Ivan M. Gubkin (1871–1939) first studied to become a teacher, then, he attended the Mining Institute in St. Petersburg. By the time of his graduation in 1910 he was almost forty years old. During the following years, he participated in geological explorations and in the two years, 1917–1918, he was a member of a delegation visiting the United States to learn about the American petroleum industry. In the 1920s, he was rapidly rising in positions in academia and beyond. From 1920, he was professor and from 1922, rector of the Moscow Mining Academy and from 1930, rector and department chair at the Gubkin Institute of Oil.

Gubkin was elected full member of the Soviet Academy of Sciences — academician — in 1929. He was 57 years old; not too young to become an academician, but unusually early if considering the scarcity of his research output. However, he and some other members of the Communist Party were elected with the obvious purpose of "Sovietizing" the Academy. In a parallel development, a large-scale repression began against members and officials of the Academy. The main thrust of this repression happened in 1929–1930, and Gubkin was an active participant in actions taken against scientists. They were sentenced to death, to exile, or to long prison terms, or they simply vanished. There were no trials; the secret police determined their fate. The repression came to end only when the authorities were satisfied that the Communist Party had taken over the Academy.

From 1931, Gubkin was also in charge of the main geological authority of the Soviet Union. His 65th birthday was widely celebrated in 1936 together with the celebration of the 40th anniversary of his scientific research activities — although he only graduated as engineer 26 years before. In the same year, he was elected vice president of the Academy of Sciences. His fame today is based on the myth created about his greatness in the late 1940s and early 1950s, during the period of anti-science terror and the triumph of Lysenkoism.

Farman Salmanov

Grave memorial (F. Rzayev) of Farman K. Salmanov and his geologist son at the Vagankovskoe Cemetery (courtesy of Julia and Keld Smedegaard).

Farman K. Salmanov (1931–2007) Azerbaijani geologist had hardship when he was growing up. His father was arrested in 1937; his mother was left with four children of which Farman was the oldest. He studied at the Azerbaijan Industrial Institute, from which he graduated in 1954, majoring in geology. He pioneered oil exploration in Siberia, which used to be considered void of oil. Salmanov's interest in Siberia originated from his grandfather who had been exiled to Siberia and who had married there before he had returned to Azerbaijan.

Salmanov began his exploration with volunteers, but without proper authorization. The authorities were considering prosecution, but had to back off because his team was so dedicated to carry on with their project. Salmanov struck oil in the Megion region in the spring of 1961. This region of Western Siberia is about three thousand kilometers (two thousand miles) east of Moscow. The official experts declared Salmanov's finding an anomaly and expected that his source would soon go dry. Instead, Salmanov found another source, then, yet another, and so on.

It was no longer possible to dismiss Salmanov and his discovery. In total, he found about 150 oil fields in Western Siberia. The oil of Siberia has become a driving force of the Soviet, then, Russian economy. Eventually, Salmanov moved to Moscow and rose in the hierarchy up to becoming the first deputy minister of oil and was elected corresponding member of the Academy of Sciences. After the collapse of the Soviet Union, he continued focusing his activities on oil exploration and worked on his projects until the end of his life.

Krylov Oil and Gas Institute

The Krylov Oil and Gas Institute and the memorial plaques of Aleksandr P. Krylov (*left*) and Mikhail L. Surguchev (*right*) on its façade; courtesy of Olga Dorofeeva.

The Krylov All-Russian Oil and Gas Scientific Research Institute at 10 Dmitrovsky Street was established in 1943 as a state institution outside of the system of the Academy of Sciences. It is still state property, but operates as a joint-stock company. It has been the principal research institute of the petroleum industry in Russia. It is also an institution of post-graduate education.

Aleksandr P. Krylov (1904–1981) was a geologist and oil explorer, a graduate of the Leningrad Mining Institute. He worked at this Institute from 1951 and was its director from 1960, holding leading positions in the Soviet oil and gas industry and research. Mikhail L. Surguchev (1928–1991) was a petroleum scientist and explorer. He worked at this Institute from 1966. During the last five years of his life, he was its director and worked on integrating scientific research and the industrial production of petroleum and its products.

University of Technology

Some of the buildings of the gigantic complex of the recently created institution of higher technological education are at the beginning of Leninsky Avenue. It is the National University of Science and Technology, comprised of a number of previously independent schools.

MISiS at 6 Leninsky Avenue — the University of Science and Technology. Its façade displays its principal areas of inquiry: Nanotechnologies; Information Technologies; Economics; and Control.

The abbreviated name, MISiS, is not spelled out, but originally it corresponds to the old name of one of the components of this new institution. It was the Moskovsky Institut Stali imeni Stalina (MISiS), which means the Stalin Moscow Institute of Steel.[2]

Row of busts in the entrance lobby of MISiS (courtesy of Olga Dorofeeva). *From left to right, top*: Pavel P. Anosov; Aleksandr A. Baikov (Chapter 1); Ivan P. Bardin (Chapter 1); Anatoly M. Bochvar (further down in this Chapter); *bottom*: Dmitry K. Chernov; Mikhail V. Lomonosov (Chapter 2); Mikhail A. Pavlov; and Aleksandr I. Tselikov (further down in this Chapter).

Pavel A. Anosov (1796–1851) was a mining engineer and metallurgist. He helped establishing heavy industry in Russia, investigated nature in the southern Ural region, and served as governor of Tomsk.

Dmitry K. Chernov (1839–1921) was a metallurgist who discovered polymorphic transformation in steel and constructed phase diagrams of iron-carbon systems. He was a professor of the Artillery Academy in St. Petersburg.

Mikhail A. Pavlov (1863–1958) was a metallurgist, active in the development of heavy industry in the Soviet Union. He contributed to the technology

[2] I. V. Stalin's original surname was Dzhugashvili. When he needed a cover-name for his clandestine operations, he chose a derivative of the Russian word *stal* meaning steel in English.

of the utilization of oxygen in metallurgy. He taught at MISiS and authored a monograph on iron metallurgy.

One of the member institutions is the Mining Institute at 6 Leninsky Avenue. The façade of its main building is distinguished by a row of sculptures.

The Mining Institue at 6 Leninsky Avenue with a row of sculptures on the top of the columns of its façade.

The Mining Institute was founded in 1918 and has changed names and affiliations several times. The eight sculptures on the façade of the Mining Institute contain twice four different figures. Although this was always an institution for training engineers, the sculptures depict miners and factory workers rather than engineers.

The new umbrella university has high standing in international listings, but it is a rather heterogeneous institution. The Nobel laureate physicist Aleksei Abrikosov (Chapter 1) used to teach in one of its former institutes. Its graduates include the crystallographer Boris Vainshtein (Chapter 1) and Dmitry V. Livanov (1967–), the former minister of education and science (2012–2016).

Vladimir Zworykin

Statue of Vladimir K. Zworykin with a television set in the Ostankino Park.

A statue in the Ostankino Park, opposite the Ostankino Television Tower across Lake Ostankinsky commemorates a Russian-born American inventor Vladimir K. Zworykin (1889–1982). Next to his figure, the memorial depicts a television set with a hole in the place of the screen through which one can view sceneries as if they were part of a TV show. Zworykin was born in the town Murom on the Oka river, some 300 kilometers (190 miles) due east from Moscow. He graduated from the St. Petersburg Polytechnic Institute as electrical engineer in 1912. Boris L. Rozing (1869–1933) was his mentor who was experimenting with cathode ray tubes to develop something of a forerunner of today's television. In 1931, Rozing was arrested and exiled and only in 1957 was he rehabilitated. Zworykin continued his studies in Paris, as a student of College de France under Paul Langevin, served in the Russian Army in World War I, and immigrated to the United States in 1919. He joined the Westinghouse Company in 1920 and continued his experiments, which led to two substantial patents in 1924 and 1929. In 1926, he earned his PhD degree from the University of Pittsburgh. In 1929, he joined the electronics company RCA (Radio Corporation of America). Its fellow-Russian immigrant head, David Sarnoff (1891–1971), was very enthusiastic about the prospects of television. Sarnoff was a pioneer of electronic communications and headed the National Broadcasting Company (NBC). Of all of Zworykin's many inventions and patents, his television has remained the most famous. Nonetheless, he always declined to be called the father of television, saying that it came about as a result of collective efforts by many.[3] Zworykin received a large number of high awards, including the recently established National Medal of Science from President Lyndon B. Johnson in 1966.

Bauman University

Bauman University is short for Bauman Moscow State Technical University. Tenderly, it is sometimes referred to as "Baumanka."

The festive gate on the left flanked by two lions is closed. The rigorously controlled entrance to the campus is through the small building on the right at 5 Second Baumanskaya Street.

Bauman University began in 1830 as an industrial school, the Moscow Crafts School. In 1868, it was made into the Imperial Moscow Technical

[3] Philo T. Farnsworth (1906–1971) was one of them and many consider him the father of television. His statue — a gift of the State of Utah — is in the National Statuary Hall Collection of the Capitol in Washington, DC.

School. Its first building from the middle of the 19th century is still there with a Minerva sculpture group on its top. Beneath the sculpture there are the insignia of three Soviet-era awards to the School.

The internal courtyard of the old section of the main building (D. Zhilyardi) with the Minerva sculpture group (I. P. Vitali) at the top.

In Soviet time, the School was renamed Moscow Higher Technical School and from 1930, its name has been associated with Nikolai E. Bauman (1873–1905), a veterinarian and a revolutionary. He was killed in a 1905 revolutionary action, not far from the School. His body lay in state at the School and his funeral procession started from there. Today, he is commemorated in several locations of the School. His most conspicuous statue dominates the square in front of the initial building.

Today, Bauman University is one of the top institutes of technology in Russia. It has had legendary figures of Russian science and technological innovation as students and as professors, such as Sergei Chaplygin, Pafnuty Chebyshev, Petr Lebedev, Dmitry Mendeleev, Vladimir Shukhov, Dmitry Sovetkin, Sergei Vavilov, and Nikolai Zhukovsky (all figure elsewhere in this book). This institution was among the first that started the military education of students. Research and development for defense purposes have been high among the priorities in its activities.

Bauman University has been very good in paying homage to its former students and professors who have earned fame for their achievements. Nikolai Zhukovsky has a number of memorials at Bauman.

The memorial plaque and bust commemorate Nikolai Zhukovsky's 50 years of lecturing on mechanics and aerodynamics, and honor him as "the father of Russian aviation."

Nikolai E. Zhukovsky (1847–1921) was a scientist of world renown in mechanics and aerodynamics. He graduated from Moscow University in 1868. He started his career as a physics teacher in a women's high school. Soon, he moved on to teaching mathematics and mechanics in what is today Bauman University. He founded the Department of Theoretical Mechanics in 1878 and remained in its charge for 43 years. He was also appointed to a professorship of applied mechanics at Moscow University in 1886.

Memorial plaque and bust honor Nikolai Zhukovsky as the founder of the Department of Theoretical Mechanics.

Zhukovsky conducted his research on a broad basis, including the investigation of how birds fly, and his work was characterized by mathematical rigor. He visited the pioneer of aviation in Germany, Otto Lilienthal. Zhukovsky created his fundamental teachings about the forces lifting airplane wings by understanding the nature of vortices. He was always ready to apply his knowledge to solving practical problems, as happened, for example, in case of the 1898 accident of the Moscow water-mains.

Zhukovsky formed student circles for advanced studies at Bauman University and outstanding future airplane designers were among his pupils. He initiated the foundation of new institutions and promoted the cause of flying. At the same time, he had some limits to his progressive thinking when he declared Albert Einstein's contributions to physics superfluous. Zhukovsky considered Galilei's and Newton's classical mechanics sufficient for solving all emerging problems in his practice.

There are other memorials of Zhukovsky beside the ones at Bauman University and Moscow University (Chapter 2). There is a Zhukovsky bust along with a Tsiolkovsky bust at the Petrovsky Palace, 40 Leningradsky Avenue.

Busts in front of the Petrovsky Palace. *Left*: Nikolai Zhukovsky (G. V. Neroda, 1959, to the left, if facing the Palace) and *right*: Konstantin Tsiolkovsky (S. D. Merkurov, 1959, to the right if facing the Palace). *Middle*: the Petrovsky Palace as viewed from Leningradsky Avenue.

The Petrovsky Palace was designed by Matvei F. Kazakov and built in 1776–1782 by the order of Catherine II. It was intended for a resting place for the tsar or tsarina when traveling from St. Petersburg to Moscow — it is close to the Moscow destination (today, within Moscow). The travelers arrived fresh and rested after this stopover. The Palace has been well preserved and it used to be the seat of the Zhukovsky Air Force Academy. Yury Gagarin, Valentina Tereshkova, Sergei Ilyushin, Aleksandr Yakovlev, Artem Mikoyan were among its famous graduates (all figure elsewhere in this book). Some time ago, the Zhukovsky Air Force Academy was joined with the former Gagarin Air Force Academy and now it is the Zhukovsky-Gagarin Air Force Academy in Moscow and Monino, Moscow Region. The Palace now functions, close to its original purpose, as a hotel for traveling dignitaries.

Chaplygin memorial (Z. M. Vilensky, 1960) at the corner of Maly Kharitonievsky and Bolshoi Kharitonievsky Streets.

Sergei A. Chaplygin (1869–1942) graduated from Moscow University in 1890 and stayed on for further training at Zhukovsky's recommendation. Chaplygin taught mechanics at Moscow University and at Bauman University along with other institutions. He served as the director of the Higher Courses for Women between 1905 and 1918, that is, until the institution was transformed into another organization, which he also headed for a short while.

When Zhukovsky founded what later became the Zhukovsky Institute of Aero-hydrodynamics, he involved Chaplygin in this project. Upon Zhukovsky's death Chaplygin became its director and he continued in this position throughout the 1920s. During the 1930s, he headed a large laboratory of aerodynamics at this Institute. A number of future leaders of Soviet aviation and space technology received their training in this Institute, among them, Mikhail Lavrentiev, Mstislav Keldysh, Nikolai Kochin, Leonid Sedov, Leonid Leibenzon, and Sergei Khristianovich. The visits of internationally renowned scientists were an indication of the high standing of the Institute. They included Ludwig Prandtl, Theodore von Kármán, Umberto Nobile, and Tullio Levi-Civita. The Pedagogical Council of the Institute had such members as Pafnuty Chebyshev, Dmitry Mendeleev, and the President of the Massachusetts Institute of Technology, John D. Runkle.

Anatoly M. Bochvar (1870–1947) was a renowned metallurgist. In 1897, he graduated from what is today Bauman University and stayed on, eventually becoming a professor. He initiated a new course, metallography. He taught at other schools as well, including the college of chemical technology that had formed in the meantime, and is today the Mendeleev University.

Memorial of Andrei A. Bochvar (N. B. Nikogosyan, 1977), at Rogov Street (https://commons.wikimedia.org/wiki/File:Памятник_Бочвару.jpg; author "Melexin;" Creative Commons; 01/11/17).

Anatoly Bochvar's son, Andrei A. Bochvar (1902–1984) followed in his father's footsteps and graduated from Bauman University in 1923. He was sent for postgraduate studies to Gottingen in 1925. From 1930, he taught at Bauman University and was head of the department his father had founded. His main research concerned the properties of metals and alloys, eventually, uranium and plutonium among them. In WWII, he worked on perfecting weaponry, including the famous T-34 tank. In 1946, he joined the nuclear program at a secret installation and worked on the production of plutonium bombs. In 1953, he returned to Moscow and served as director of what is today the Bochvar High-technology Research Institute for Inorganic Materials, at 5 Rogov Street.

We mention here a few more Bauman graduates and professors and present a few memorial plaques on the Bauman campus. Some of these scientists appear also elsewhere in this book as do other Bauman graduates.

Memorial plaque of Dmitry K. Sovetkin (courtesy of Artemy Maslov).

Dmitry K. Sovetkin (1834–1912) was a mechanical engineer, one of the early graduates of the School. He is considered to be the founder of the Russian method of teaching industrial arts and crafts.

Nikolai I. Mersalov (1866–1948) was a mechanical engineer who graduated from Moscow University and worked as an engineer in Germany for some years. He earned his higher qualifications from Bauman University. In 1898, he founded a department of thermal physics at Bauman University. He taught also at Moscow University and at institutions of higher education for agriculture.

Vladimir V. Uvarov (1899–1977) was a Bauman graduate and spent his entire professional life at his Alma Mater. He earned higher qualifications from Moscow University. In 1949, he founded a department for gas turbines and non-traditional energy apparatus at Bauman University.

Anatoly I. Zimin (1895–1974) pioneered research in metallurgy utilizing high pressure for which he founded a department at Bauman University.

Memorial plaques of Georgy A. Nikolaev (*left*) and Aleksei M. Isaev (*right*); courtesy of Artemy Maslov.

Georgy A. Nikolaev (1903–1992) graduated from Moscow University and was a specialist in welding. He started his career at Bauman in 1930 and headed the department of machines and the processes of welding between 1947 and 1989, serving also as rector between 1964 and 1985. He participated in the construction of many famous structures in Moscow and elsewhere.

Aleksei M. Isaev (1908–1971) was a rocket engineer who designed efficient smaller anti-missile and anti-aircraft rockets, including the "Scud" rockets. The Scud name was coined by Western intelligence agencies; these rockets came in a wide variety, and were mostly deployed in countries friendly with the Soviet Union during the Cold War.

Left: Vladimir P. Barmin was a specialist in designing structures for rocket launch. *Right*: Evgeny A. Chudakov (Chapter 1) was a specialist in machine construction.

Vladimir P. Barmin (1909–1993) was a designer of jet and rocket engines and one of the founders of Soviet cosmonautics. In 1936, he visited the United States. In WWII, he worked on the perfection of the "Katyushas." After WWII, Barmin was appointed to be one of the leaders of the Soviet efforts in creating the rocket technology for the armed forces. He was a member of the informal council of chief designers headed by Sergei Korolev. Barmin founded and headed the department of launch installations of rocket facilities at Bauman University and he was also in charge of developing automated structures for the exploration of the Moon and Venus.

Memorial plaques of two former students, Sergei P. Korolev (*left*), between 1926 and 1930 and Nikolai A. Pilyugin (*right*), between 1931 and 1935 (more about both, below).

Sergei Korolev has a number of other memorials on the Bauman campus. All have appeared after his death. While he was alive, his identity and activities were strictly classified in order to protect him. When he passed away, he became a celebrity overnight (more about him later in this chapter).

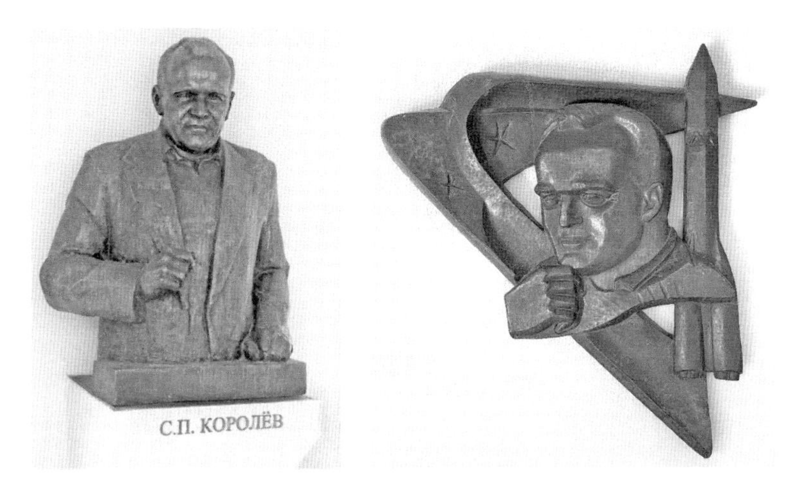

Two of the Korolev memorials on the Bauman campus. The image on the right is courtesy of Artemy Maslov.

This statue depicts Korolev with the first Sputnik and rockets. This image (courtesy of Artemy Maslov) shows the authors, second and fourth from the right, in the company of our hosts.

Memorial plaque of the aircraft designer Pavel O. Sukhoi on the Bauman campus (*left*) and on the façade of the building where he used to live, 7 Leninsky Avenue (*right*).

Pavel O. Sukhoi (1895–1975) originated from the Vitebsk Region in today's Belarus. He was a member of Zhukovsky's student circle for advanced studies of aviation. WWI, the revolutions, and the Civil War interrupted Sukhoi's studies and he graduated only in 1925. By then Zhukovsky had died and Andrei Tupolev was Sukhoi's mentor. Sukhoi continued working with Tupolev at the Zhukovsky Institute of Aero-hydrodynamics, and he became a prolific designer of airplanes that participated in WWII. He was among the first Soviet engineers designing jets. The plant where he worked as director of the construction bureau is now the Sukhoi Experimental Construction Bureau (23 Polikarpov Street).

Left: Eduard Satel was a Bauman graduate and professor. *Right*: The aircraft designer Andrei Tupolev was one of the most famous Bauman graduates.

Eduard A. Satel (1885–1968) majored in machine construction and held responsible industrial appointments; among them, he directed the establishment of the Stalingrad Tractor Plant, the first Soviet plant with an assembly line. He initiated the Stalingrad Institute of Mechanics. In 1937, he was appointed professor at Bauman University and he organized and headed a department of special technologies. In WWII, he worked on the transformation of civilian industry to the production of arms. Following WWII, he worked on the scientific foundations of mechanical engineering.

Andrei N. Tupolev (1888–1972) had an unusual career although not uncommon among successful Soviet scientists and technologists. He began as a prolific aircraft designer, but in 1937 he was arrested upon trumped-up charges during Stalin's first great terror wave. Some of his colleagues were executed; Tupolev was sentenced to ten years prison and survived. Soon he was transferred to a special labor camp where specialists were brought together to design aircraft. This was one of the so-called "sharashkas" where "enemies of the people" worked as slave laborers, but where the conditions were better than in ordinary prison camps. He was rehabilitated only after Stalin's death. Tupolev not only designed aircraft, he was also engaged in reverse engineering. He replicated the American B-29 bomber and called it Tu-4. Tupolev's best known commercial airplanes were among the first jet airliners, the Tu-104 and the Tu-154. Tupolev's son, Aleksei A. Tupolev (1925–2001) designed the supersonic T-144.

Top: The main building on the bank of the Yauza river. *Bottom*: Front view (*left*) and side view (*right*) of the row of sculptures on top of the colonnade.

Another well-known building of the Bauman campus from the post-WWII period is directly on the bank of the Yauza river with a row of statues on the top of its colonnade. There is heavy emphasis on manual workers among the six sculpted figures; some of them appear more like steelworkers, foundry men, or blacksmiths rather than engineers or designers. In front of this building, on the side of the Yauza river, there are two busts facing each other.

Busts of Aleksandr I. Tselikov (*left*, courtesy of Artemy Maslov) and Vladimir N. Chelomei (*middle*, V. A. Sonin, 1984) and a memorial plaque of Chelomei (*right*, courtesy of Artemy Maslov).

Aleksandr I. Tselikov (1904–1984), a prominent metallurgist, graduated from Bauman University in 1928 as a mechanical engineer specializing in the construction of rolling machines. Parallel to industrial appointments, he was also professor of mechanical engineering at Bauman University.

Vladimir N. Chelomei was also a professor at Bauman University and there is more about him below, in the section of space exploration. The memorial plaque commemorates Chelomei specifically as the founder of the department of aerospace systems.

There is a rich collection of memorials in the Museum of Bauman University, but the memorials are not limited to the Museum. Everywhere on the campus, displays refer to the history and achievements of the University.

The exterior and a model interior of a spacecraft that returned to Earth in 2012 showing one cosmonaut. The actual spacecraft was the Soyuz TMA-04M with two Russian cosmonauts and one American astronaut onboard. The flight lasted 124 days 23 hours 51 minutes and 30 seconds.

The Museum at the University exhibits a rich collection of memorials. They demonstrate that the activities of Bauman University have closely intertwined with the industrialization of the country, the space program, and defense.

Four busts in the Museum of Bauman University (all by L.E. Kerbel). *From left to right*: Zhukovsky, Shukhov, Tupolev, and Korolev.

Moscow Energy Institute

In 1930, new institutes were created by separating some areas from Bauman University. One of them was the Moscow Energy Institute.

The main building of the Moscow Energy Institute at 14 Krasnokazarmennaya Street (http://dic.academic.ru/pictures/wiki/files/77/Moscow_Power_Engineering_Institute_main_Building_1.jpg; 01/02/17).

The current official name of the Moscow Energy Institute is Moscow Power Engineering Institute — a national research university. There have been some renowned personalities among the past staff of this School, such as Andrei Sakharov and Sergei Vavilov (both in Chapter 1). Boris Chertok (Chapter 7) and Vladimir Veksler (Chapter 1) were among its graduates.

Memorial plaques of two former professors on the façade of the main building, Karl Krug (*left*, http://dic.academic.ru/pictures/wiki/files/77/Moscow_Power_Engineering_Institute_main.jpg; author: "Solver55ris;" 01/02/17) and Vladimir Kirillin (*right*, courtesy of Olga Dorofeeva).

Karl A. Krug (1873–1952) studied first at what is today Bauman University and obtained his higher scientific qualifications in Germany. He founded the training of electrical engineers at his Alma Mater and continued his career at the Moscow Energy Institute when it separated from Bauman.

Vladimir A. Kirillin (1913–1999) was deputy prime minister of the Soviet Union between 1965 and 1980 and was responsible for science and technology. He graduated from the Moscow Energy Institute in 1936. He rose rapidly in the party and government organizations while acquiring his scientific degrees and membership of the Academy of Sciences and staying connected with thermal physics and thermodynamics. There is now a Kirillin Department of Engineering Thermal Physics in the Institute.

Memorial plaque of Sergei A. Lebedev on the façade of his former home at 21 Novopeschannaya Street (courtesy of Olga Dorofeeva).

Sergei A. Lebedev (1902–1974) was a graduate of Bauman University and a pupil of Professor Krug in electrical engineering. Lebedev's initial projects were related to the production and utilization of electrical energy. Later, he was a professor at the Moscow Energy Institute. During WWII he was involved with defense-related work. After the war, he worked in computational technology and pioneered the design of electronic computers in the Soviet Union. He remained in the forefront of creating ever improved computers throughout his career. He has been credited with a decisive contribution to creating and maintaining a balance between the American and Soviet systems of anti-ballistic missiles. From 1952, Lebedev was director of what is today the Sergei Lebedev Institute of Precise Mechanics and Computer Engineering (51 Leninsky Avenue). He had to resign from this position in the early 1970s on account of his debilitating illness, but continued working at home until the end.

Polytechnical Museum

Many examples of the achievements of Russian and Soviet engineers and technologists as well as their international counterparts are collected in the exhibition halls of the Polytechnical Museum at 3/4 Novaya Square (https://upload.wikimedia.org/wikipedia/commons/c/ce/Музей_Политехнический.JPG; Creative Commons; Author "Alexor;" 01/15/16). The Museum was founded in 1872 and the central part of its building opened in 1877 (I. A. Monihetti was the architect). The building was completed in 1907. Currently (2017) the Museum is undergoing a major renovation.

Explorers of Space

On October 4, 1957, the first artificial satellite of the Earth, Sputnik-1 (Companion-1), was launched. It was a device of spherical shape of a diameter of 580 millimeters and of mass of 83.6 kilograms. Its shortest distance from Earth (perigee of its trajectory) was 228 kilometers and its largest distance (apogee) was 947 kilometers.

From left to right: Memorial of Sputnik-1 and its creators (S. Ya. Kovner, 1958) at the metro station Rizhskaya (courtesy of Olga Dorofeeva). Vostok spaceship memorial in Ostankino Park (courtesy of Olga Dorofeeva). Gagarin memorial (Yu. L. Chernov, 1985) at 14 Vernadsky Avenue (courtesy of Vasily Ptushenko).

Memorial of Yury Gagarin, 43 meters tall, in Gagarin Square.

The Soviet achievements in space exploration were at the pinnacle of the success stories of the Soviet Union. They were meant to prove the superiority of Soviet science and technology over those of the capitalist world. Their popularity was enhanced because they had not been associated with destruction — unlike the nuclear program — although they had grown out of military preparations. The conquest of Space had enthused large masses of people all over the world and the Soviet propaganda made the most of it. Memorials carved in stone and issued as postage stamps abound. The Soviets coined the terms "cosmonauts" and "cosmonautics" instead of astronauts and astronautics, etc. They welcomed back the cosmonauts upon their return from their space travels with parades along Leninsky Avenue and on the Red Square; showered them with perks, awards, and titles; and buried their ashes in the Kremlin Wall when catastrophes killed them.

Yury A. Gagarin (1934–1968), aviator and cosmonaut, was the first human in space. He orbited our planet on April 12, 1961, in the spaceship Vostok-1 (East-1). Gagarin became a world celebrity and April 12 has become the Day of Aviation and Cosmonautics. The 108-minute flight changed

Gagarin's life and the ensuing celebrations took their toll. Nonetheless, he completed his studies at the Zhukovsky Air Force Academy and in 1966 he returned to training for future flights. He was selected to be Vladimir Komarov's backup on the first flight with the new space ship Soyuz. The flight ended with tragedy and Komarov perished.

Gagarin continued his training, but he was killed on board a MIG plane on March 27, 1968, together with an experienced flight instructor, V. S. Seregin. The bodies of Gagarin and Seregin were cremated and interred in the Kremlin Wall (Chapter 8). A state commission investigated the catastrophe and the findings of the investigation were compiled in a multi-volume classified report. Eventually, declassified information suggested that an unauthorized supersonic jet flew so close to Gagarin and Seregin's MIG that it became uncontrollable; it took a vertical turn and hit the earth. There are many Gagarin memorials in Moscow, in Russia, and internationally, too.

Between the first sputnik and Gagarin's flight, there were other flights including some that carried one or two dogs. Right after Sputnik-1, the next flight was Sputnik-2 on November 3, 1957. This time the Vostok space ship carried a two-year old dog, Laika, in addition to scientific instruments. The weight of this satellite was 508.3 kilograms. Laika survived a few revolutions around the Earth, but perished under the overheated conditions of the small cabin. Sputnik-2, with Laika's body, continued its revolutions about the Earth for 162 days and burned up in the atmosphere on April 14, 1958. Laika's participation in the space program became a world sensation.

The Laika memorial in a fenced-off yard where Militseisky Passage runs into Petrovsko-Razumovskaya Promenade, close to the Dynamo metro station.

There was strong criticism of the Soviet space program for Laika's flight was designed without an option for its safe return. Those in charge of the Soviet space program understood that there should be no more flights in which living creatures would be launched without means of their safe return. Laika's memorial is moving, but it is not near the main memorial assembly of Cosmonautics (see below). The impression is that being a reminder of an unsuccessful story, it is better for it staying in the shadows.

Laika had been carefully selected and prepared for the flight by the military physician Oleg G. Gazenko, who was responsible for the animals that were to become space travelers. Eventually, it was recognized that a research venue was needed to investigate the physiological aspects of space exploration and the Institute of Cosmic Biology and Medicine was organized in 1963. Soon the name of the Institute was changed into the Institute of Biomedical Problems as it is known today, at 76 Khoroshevskoe Highway. The combined staff of the Institute consisted of military health experts and civilian scientists of physiology and biophysics. From 2001, the Institute has been part of the system of research institutes of the Academy of Sciences.

Memorial plaques of the first three directors of the Institute of Biomedical Problems on its façade: Andrei V. Lebedinsky, Vasily V. Parin, and Oleg G. Gazenko (courtesy of Oleg V. Voloshin).

Andrei V. Lebedinsky (1902–1965) was a physiologist and biophysicist, and general of the military health service. He graduated from the Military Medical Academy in Leningrad in 1924; worked at the same Academy, then, directed a biophysical research institute. From 1963, he was the first director of what is today the Institute of Biomedical Problems. His research interest was in the physiology of vision and he investigated the impact of ionizing radiation and of ultrasound on living organisms.

Vasily V. Parin (1903–1971) was a physiologist and researcher of the cardiovascular system. Toward the end of his research activities he directed studies in cosmic medicine. He had a checkered career. He was director of the First Moscow Medical Institute (Chapter 5) in 1941–1943. According to some sources, in October 1941, ignoring his responsibilities, he ran away fearing the German occupation. He was deputy minister of health of the Soviet Union between 1942 and 1945. Various authors praise him for having prevented any epidemic during the war both at the military front and the home front. After the war he was sent for a study tour of four months to the United States. Soon after his return, in 1947, he was arrested as an American spy, freed only in 1953, and rehabilitated in 1955 (see also Chapter 5). During the rest of his life he occupied important functions in the medical establishment. In particular, he was the second director of the Institute of Biomedical Problems and he is credited with having founded cosmic cardiology.

Oleg G. Gazenko (1918–2007) was a physiologist and a general of the military health service. He graduated from the Second Moscow Medical Institute (Chapter 5) in 1941. Together with his whole graduating class, he left for the front right upon graduation. After the war he continued his training and specialized in aviation medicine. He was involved with the space program and was one of the founders of space biology and medicine. In 1960, after two dogs, Zhulka and Zhemchuzhina, returned safely from a space flight, Gazenko took in Zhulka, and the dog lived with him for another 14 years. Gazenko directed the Institute of Biomedical Problems between 1969 and 1988.

The monument "Explorers of Space" (by architects A. N. Kolchin, M. O. Barshch, engineer L. Shchipakin, and sculptor A. P. Faidysh-Krandievsky, 1964) and the Tsiolkovsky statue at its front.

The monument "Explorers of Space" in Ostankino Park is on a grandiose scale. Its principal component of 99 meters represents the rocket exhaust trail, topped by an 11 meter rocket pointing to the sky.

View of "Explorers of Space" from the start of the Promenade of Cosmonauts. The two spheres in the front represent the Globe (*left*) and the Universe (*right*).

The enormity of the monument complex can be perceived by looking at the Promenade of Cosmonauts with the "Explorers of Space" at its end. It starts with two spheres depicting the Globe and the Universe. Along the way there are a number of memorials honoring various events and participants in the history of space exploration.

A sundial representing the Sun with the planets around it. The planets around the Sun are placed in their relative positions as they were at the moment of Gagarin's flight.

Busts of four Muscovite cosmonauts that have each earned two awards of Hero of the Soviet Union. *From left to right*: Valentin V. Lebedev, Svetlana E. Savitskaya, Aleksandr P. Aleksandrov, and Vladimir A. Soloviev. In 1984 Savitskaya was the first woman to walk in Space.

Busts of five cosmonauts of special credentials: *from left to right*: Yury A. Gagarin (L. E. Kerbel, 1967), Valentina V. Tereshkova (1937– ; first woman in space), Pavel I. Belyaev (1925–1970; commander of the first mission of space walk in 1965; A. P. Faidysh-Krandievsky, 1967), Aleksei A. Leonov (1934– ; first space walk; A. P. Faidysh-Krandievsky, 1967), and Vladimir M. Komarov (1927–1967; successful space flight with Voskhod-1, but perished with Soyuz-1 in 1967).

From across the square from the row of the busts of the five cosmonauts, there are statues of scientists and engineers that were the principal contributors to the Soviet success in space exploration.

Sergei P. Korolev, leader of the rocket-cosmonautics program (S. A. and S. S. Shcherbakov, 2008).

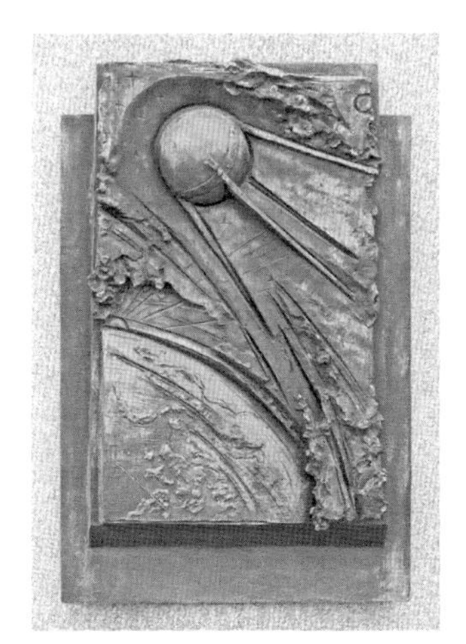

Side panels of the Korolev statue represent (*from left to right*) the first Sputnik, a scene of rocket launching in Baikonour, Kazakhstan, and the first space walk by Aleksei Leonov.

Sergei P. Korolev (1906–1966) was in charge of the development of the Soviet rocket-cosmonautics system. When he was three, his mother left the family, but he was reunited with her at the age of ten. By then he was a junior high school student in Odessa. His stepfather had training in engineering, which helped Korolev develop his interest in all things technical and in aviation in particular. Korolev began his college education in the Kiev Polytechnic Institute, then moved to Moscow and continued at Bauman University under Tupolev's mentorship. Following a meeting with Konstantin Tsiolkovsky, Korolev became a lifelong enthusiast of rocket technology. Korolev and his colleagues built devices, but Stalin's terror interrupted this development. Korolev was arrested in 1938 on trumped up charges and underwent rough interrogations. According to unconfirmed descriptions, this included psychological abuse and torture during which both his jaws might have been broken. In 1940, Korolev was sentenced to eight years of incarceration.

Korolev worked in a special camp, again under Tupolev who in the meantime had also been arrested and sentenced to a prison term. Korolev successfully designed airplanes and torpedoes that proved useful for the military. He was then moved to another special prison where he worked on rocket motors. Korolev was freed from prison in 1944, but was not rehabilitated. From this time Korolev was in top positions in the projects of the Soviet rocket technology, including the projects of ballistic missiles. He directed the program that eventually included automated satellites and piloted spaceships. There were a number of chief designers in the Soviet rocket technology and Korolev was not just another one; he chaired the council of chief designers. Vladimir Barmin, Aleksei Bogomolov, Valentin Glushko, Viktor Kuznetsov, Nikolai Pilyugin, and Mikhail Ryazansky were among the other members of the council.

Korolev maintained a close personal relationship with the cosmonauts. The public celebration of the cosmonauts by the leaders of the Soviet Union was his triumph too, but he was kept in the shadows until after he died. A lot has been written about Korolev's illness and the surgery that was not supposed to be life threatening, but which he did not survive. He was operated on by the Minister of Health of the Soviet Union. Complications arose during the operation and some of the medical means that should have been employed were hindered by the consequences of traumas Korolev had suffered in his interrogations.

His peers remembered Korolev as without hope at the time of his repression. He kept saying that he would perish without even an obituary. What happened after his death could not have been more different. The accolades and honors were not only in stark contrast with his prison life, but also with his obscurity during his active life. His resting place is in the Kremlin Wall.

Sergei P. Korolev's home from 1959. His statue stands in the garden.

Right after the first spectacular successes of the space program, the Soviet government offered Korolev and each of the other leaders of the program a comfortable apartment in Moscow and a weekend house (*dacha*) in the outskirts. Instead, Korolev asked for a comfortable house in Moscow. It was built according to his specifications; its style give the impression of a Bauhaus afterthought, and in the West, even in a big city, such a house could have belonged to an upper-middle-class professional. There was though a difference. Korolev's house was built on a huge lot of prime real estate, and it was guarded day and night by a large team of security agents.

Today, there is a memorial museum in Korolev's house at 28 First Ostankinskaya Street. The lot around the house is much reduced, but it is still fenced off and guarded. According to our experience, even during the opening hours of the museum, the gate is closed and opens only when a visitor declares through the Intercom about his/her desire to visit the museum.

From left to right: Statues of Vladimir N. Chelomei, Valentin P. Glushko, and Mstislav V. Keldysh (Yu. L. Chernov, 1981) next to the monument "Explorers of Space."

Vladimir N. Chelomei (1914–1984) graduated from the Kiev Aviation Institute in 1937 and worked in Moscow from 1941. Along with his leading positions in the development of the Soviet rocket technology, he did research in mechanics and automation and was professor at Bauman University.

Valentin P. Glushko (1908–1989) was interested in rocketry from early youth, designed aircraft, and corresponded with Tsiolkovsky. He graduated from Leningrad University and he designed an interplanetary spaceship for his thesis. His career was interrupted by his arrest in 1938 and incarceration, and he worked as an aircraft designer in a special prison. He was freed in 1944, worked on rocket fuels and engines, and became one of the leaders of the space program. There is a Glushko memorial plaque on the façade of his former home between 1955 and 1965, at 1 Kudrinskaya Square, in downtown Moscow.

Mstislav V. Keldysh (1911–1978) graduated in mathematics from Moscow University in 1931 under Nikolai Luzin's mentorship. Keldysh began his research career in the Zhukovsky Institute. From 1934, he did his research at the Steklov Institute in Mikhail Lavrentiev's group. From 1953, until the end of his life, he directed what is now the Keldysh Institute of Applied Mathematics. Keldysh was also a professor at Moscow University and an organizer of science, serving as President of the Academy of Sciences from 1961 until 1975.

Thematically, though not geographically, the monument of Nikolai A. Pilyugin (1908–1982) belongs here. He was an engineer designer and member of the informal council of chief designers of the space program. He was a Bauman graduate, and his main interest was in automation and automatic control of rockets and other systems of cosmonautics.

Left: Statue of Nikolai A. Pilyugin (V. B. Soskiev, 2008) in Academician Pilyugin Street where Pilyugin Street and Architect Vlasov Street intersect. *Right*: Bust of Viktor I. Kuznetsov in a small park in Aviamotornaya Street (courtesy of Olga Dorofeeva).

Viktor I. Kuznetsov (1913–1991) was a specialist in applied mechanics and automation in rocket and space technology. He graduated from an industrial college in Leningrad in 1938. He excelled in building scientific instruments and learned the theoretical foundations of mechanics. He joined the nuclear program, but soon he transferred to the rocket and space program and became one of its principal designers.

Statue of Konstantin E. Tsiolkovsky (A. P. Faidysh-Krandievsky) in front of "Explorers of Space."

Konstantin E. Tsiolkovsky (1857–1935) originated from Polish nobility in Western Imperial Russia and lived the last four decades of his life in Kaluga, about 160 kilometers (a hundred miles) southwest from Moscow. Deafness and family tragedies exacerbated the hardship of his childhood. He studied on his own and visited a library in Moscow as diligently as if it were his most rigorous work place. He then earned a license to teach science. His first research papers showed inventiveness but also the disadvantage of isolation from the scientific world, although he had some interactions with such giants of science as Mendeleev, Sechenov, and Stoletov.

Tsiolkovsky was full of ideas in diverse areas, beside his research of Space. He proposed, for example, a construction for a dirigible and wrote science fiction. In his experiments, he used mostly self-constructed instruments. After the 1917 revolution, the new, the so-called Communist Academy accepted him with a life-time pension, but the Academy of Sciences never voted him into its membership. Immediately upon Tsiolkovsky's death the Soviet State declared his collected works a national treasure. In Russia, he is revered as the founder of cosmonautics.

Memorial plaques of Mikhail S. Ryazansky (*left*) and Leonid I. Gusev (*right*) on the façade of the corporation "Russian Space System," 53 Aviamotornaya Street (courtesy of Olga Dorofeeva).

The corporation known as the Russian Space System has a more elaborate name, something like the Russian corporation for rocket and space instrumentation and information systems. It is a conglomerate of several organizations and institutions dealing with rocket and space technology. One of its predecessor institutions was founded by the Soviet government in 1946. On the façade of its headquarters there are two memorial plaques.

Mikhail S. Ryazansky (1909–1987) worked in plants and excelled in the amateur radio movement for years before he graduated from the Moscow Energy Institute in 1935. During WWII, he contributed to the development of the Soviet radio locator system. He was in the group of Soviet specialists who in 1945–1946 investigated the V-rocket system in Germany. Subsequently, he participated in creating the Soviet rocket and space technology along with Korolev, Glushko, and others. He occupied leading positions in the project and was one of the members of Korolev's informal council of chief designers.

Leonid I. Gusev (1922–2015) was a leading authority in space and defense technology. He was a worker and fought in WWII, and continued his studies afterwards. He graduated in 1955 from what is now the Moscow Technical University of Communication and Informatics (8 Aviamotornaya Street). He rose in his positions while continuing his activities in radio technology. He served as deputy minister of the ministry called general machine building, responsible for rocket and nuclear armaments. Gusev was in charge of what is today the "Russian Space System" for the period of 1965–2004.

Museum of Cosmonautics

The Memorial Museum of Cosmonautics is beneath the monument "Explorers of Space," at 111 Mir Avenue, which is part of the Promenade of Cosmonauts.

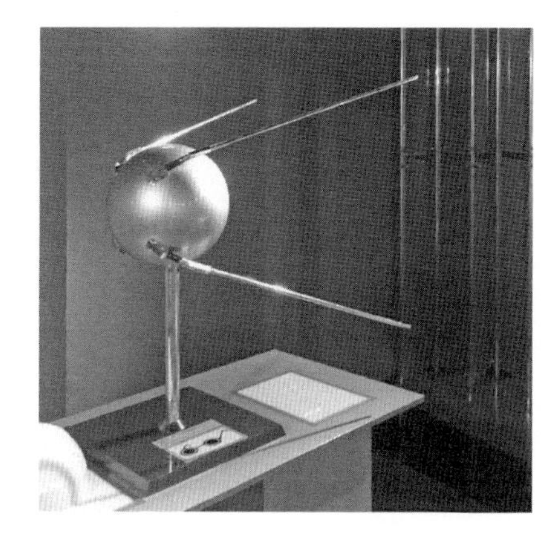

Models of Sputnik-1 (*left*) and Vostok-1 (*right*).

Taxidermy made it possible to see Belka and Strelka on display at the Museum.

Belka (Squirrel or Whitey) and Strelka (Little Arrow) were among the passengers of the Vostok-type spaceship on August 19, 1960. The other passengers included a grey rabbit, mice, rats, flies, and plants. It was for the first time that living creatures went into orbit and returned alive. Belka and Strelka became instant international celebrities. Their flight, and other, similar flights, paved the way for Gagarin's space flight in 1961.

Spacewalk

Landing and waiting to be discovered.

The demonstration of the Apollo-Soyuz Test Project in 1975 in the National Air and Space Museum in Washington, DC. The command and service modules of an Apollo spacecraft (*left*) were linked by a docking module to a Soyuz spacecraft (*right*).

The American counterpart of the Museum of Cosmonautics is the National Air and Space Museum of the Smithsonian Institution in Washington, DC. There is plenty of recognition of the Soviet successes there.

Display of photographs of Soviet cosmonauts and spacecraft models in the Museum of Cosmonautics.

Two sources of Soviet pride in space exploration modelled in the Museum of Cosmonautics; *left*: Konstantin Tsiolkovsky in his workshop and *right*: Sergei Korolev in his office.

From the collection of reliefs in the entrance lobby of the Museum depicting real and mythical scenes and devices of historic importance for conquering Space.

From left to write: Reliefs of Copernicus, Giordano Bruno, Galilei, Kepler, Newton, Lomonosov, and Einstein at the start of the exhibition.

The internationalism of this collection is remarkable. Placing Mikhail V. Lomonosov among Newton and Einstein along with the rest of this set of giants may be questionable from the point of view of original scientific contribution. However, his position in Russian science history and the aura that has developed about him justifies the presence of Lomonosov's image here.

Statue of Nil F. Filatov (V. E. Tsigal, 1960), the founder of pediatrics in Russia, Bolshaya Pirogovskaya Street at the southern corner of the Deviche Pole Park.

5

Medical City

Introduction

We first visited the medical campus of Sechenov Medical University around the Bolshaya Pirogovskaya Street after one of our visits to the Novodeviche Cemetery. The area and hence the campus is called Deviche Pole. The Cemetery is on Luzhnetsky Street and we continued walking along it in the north-eastern direction, toward the center of the city. Luzhnetsky Street becomes Bolshaya Pirogovskaya Street after the Novodeviche Street and we were already on the medical campus.[1] There are statues, busts, and memorial plaques honoring renowned physicians along the streets and within the walkways and parks of the campus.

The Sechenov Medical University is one of the two principal medical schools in Moscow. The other is the Pirogov Medical University. There may be a slight confusion because the Sechenov Medical University is around the Bolshaya Pirogovskaya Street and the Malaya Pirogovskaya Street, which runs parallel to it. The Pirogov Medical University is far from this area, in Ostrovityanov Street, near the intersection of Leninsky Avenue and Ostrovityanov Street. We start with a brief sketch of a few important milestones in the histories of these schools.

The full name of the Sechenov Medical University is I. M. Sechenov First Moscow State Medical University. It traces its origin back to the medical school of Moscow University founded in 1758. Many renowned physicians have graduated from this School, such as Sergei Botkin, Aleksandr Myasnikov, Nikolai Pirogov, Ivan Sechenov, Nikolai Sklifosovsky, and the famous writer and playwright Anton Chekhov who was a physician by training.

[1] Most of the campus is between Bolshaya Pirogovskaya Street and Pogodinskaya Street and the numbering is descending, from No. 6 to No. 2, on the Bolshaya Pirogovskaya Street if starting from the south-west. Several buildings may belong to the same number and the buildings on the eastern side of Pogodinskaya Street are part of the Bolshaya Pirogovskaya addresses.

Statue of Anton P. Chekhov (A. I. Rukavishnikov, 2014) near the Medical Center of Moscow University (courtesy of Olga Dorofeeva). There is another Chekhov statue (M. Anikushin, 1998) in Kamergersky Street, near the Chekhov Arts Theater in downtown Moscow.

Moscow University had the first lectures in medicine in 1758 and continuous instruction in medicine began in 1764. The University was granted the rights to confer the degree Doctor of Medicine in 1791. It opened the first teaching hospital on Nikitskaya Street in 1805 and it started publishing a medical periodical in 1806. The Deviche Pole campus with its teaching hospitals and research institutes was constructed during the decade of 1887 to 1897. In 1925, a postgraduate course was established and in 1930, the medical school of Moscow University became an independent institution, the First Moscow Medical Institute, which was named after Ivan Sechenov in 1955. It acquired its current name in 2010.

The former main building (S. U. Soloviev, 1912) of the Higher Courses for Women at the intersection of Malaya Pirogovskaya, Rossolimo, and Kholzunov Streets. Currently, it belongs to the Moscow Pedagogical University.

Reliefs on the façade illustrate some of the activities of the Higher Courses for Women.

The other major medical school, the Pirogov Medical University, traces its origin back to the Higher Courses for Women. There, the first lectures in medicine were delivered in 1906, with the first graduates receiving their diplomas in 1912. The revolutions in 1917 initiated sweeping changes in the organization of this institution and the separation of sexes in higher education was soon abolished. In 1918 the Higher Courses for Women was transformed into the co-educational Second Moscow University. In 1930, this medical school also became independent and the Second Moscow Medical Institute was established. In 1946, the Second Moscow Medical Institute was named after Stalin. The next change in name came in 1957, and the new name was Pirogov Moscow State Medical Institute. This became in 1991 the N. I. Pirogov Russian State Medical University, and now it is the Pirogov Russian National Research Medical University — in short, Pirogov Medical University.

The labels of First and Second refer to the time when these medical schools started functioning, that is, 1758 and 1906, respectively. There is then another medical school, the Moscow State Medical University of Stomatology, which started in 1922. In 2012, it was named after the founder of dentistry in the Soviet Union, the A. I. Evdokimov Moscow State Medical University of Stomatology. For a while, there used to be a Third and even a Fourth Medical Institute for training physicians in Moscow. The Third was established in the early 1930s on the basis of the Moscow City Hospital and the Fourth in 1941 on the basis of the Moscow Regional Hospital. Both were evacuated during the war. Upon return, the two united in 1943, and that was the Moscow Medinstitute. This Institute was transferred to the city of Ryazan, some 170 kilometers (one hundred miles) southeast from Moscow; it is now the I. P. Pavlov Ryazan State Medical University.

Numerous memorials related to medicine are scattered about Moscow, mainly on the campuses of medical institutions. There are then memorials to medical personnel that helped on the battlefield and fell in action and also to high-ranking generals of the medical services of the armed forces.

Under the Soviet system, the civilian medical profession was not in very high esteem by the authorities as it was not considered to be a "forces of production." Paradoxically, the conditions of neglect, compounded by the isolation from the international scene, led to some original discoveries and valuable innovations in selected areas. Suffice it to mention the pioneering contributions of the transplant pioneer Vladimir Demikhov and the ophthalmologist Svyatoslav Fedorov.

Our presentation starts with the memorials of the oldest department of physiology in Russia and continues with the memorials at the Deviche Pole campus. Then, a selection of memorials elsewhere in Moscow follows.

Department of Normal Physiology

The Department of Normal Physiology of Sechenov Medical University and Ivan M. Sechenov's memorial plaque on its façade (courtesy of Olga Dorofeeva).

The origin of the Department of Normal Physiology, at 11 Mokhovaya Street, dates back to 1776, shortly following the establishment of medical training at Moscow University. Great physicians served as heads of department and as professors of physiology at this institution. At the time of Ivan M. Sechenov's leadership between 1891 and 1901, the department was often referred to as *the Institute* of Physiology.

Memorial plaques of Mikhail N. Shaternikov, Ivan P. Razenkov, Petr K. Anokhin, and Konstantin V. Sudakov (courtesy of Olga Dorofeeva).

In addition to Sechenov's memorial plaque, there are four more plaques on the façade of the building.

Mikhail N. Shaternikov (1870–1939) was a professor of physiology and Sechenov's close associate. Shaternikov often constructed his own experimental apparatus for his research. In addition to physiological investigation he was also dedicated to establishing guidelines for a healthy diet for different age groups of the population.

Ivan P. Razenkov (1888–1954) was a leading physiologist in the Soviet Union. He was professor of medicine at several institutions of higher education, edited medical journals, and was president of the All-Union Society of Physiologists. He directed an academic research facility of experimental medicine in Moscow and from 1939, he was the chair of the Department of Normal Physiology. In 1951, his career was abruptly over when he was accused of ostensibly advocating anti-Pavlovian ideas in physiology.

Petr K. Anokhin (1898–1974) was in charge of the Department of Normal Physiology between 1955 and 1974. He was one of Ivan P. Pavlov's disciples. Anokhin's main interest was in the fundamental investigation of life itself and in another area, which was close to what we call today *systems biology*. He was also a leading specialist in neurological surgery. In 1950, a group of physicians accused Anokhin with having strayed from Pavlov's teachings. Anokhin was dismissed from his positions and his story could have ended even worse. After Stalin's death, however, he regained his high positions in the hierarchy of Soviet medicine.

Konstantin V. Sudakov (1932–2013) led the Department between 1974 and 2013. He expanded Anokhin's studies of functional physiology and initiated the Anokhin Research Institute of Normal Physiology of the Academy of Sciences at the same venue where the Department is. In 2013, Sudakov gave up his chairmanship and continued as honorary chairman, but died shortly after his resignation.

Deviche Pole Campus

The first building on the left along Bolshaya Pirogovskaya Street, is the front of the University Clinical Hospital No 1. The big complex of clinics includes the Ostroumov Clinic of Internal Medicine; the clinic of ear, throat, and nose; the Vinogradov Clinic of Internal Medicine; the Burdenko Clinic of Surgery; the clinic of traumatology, orthopedics, and pathology of joints; and the clinic of cardiovascular and general surgery. Just before reaching the short Abrikosov Street, there is a church named after the holy Dmitry Prilutsky, which used to be a medical building under the Soviet system (see below).

Abrikosov Street

Left: Bust of Aleksei Abrikosov (A. G. Postol, 1960) and *right*: Bust of Boris Petrovsky (courtesy of Olga Dorofeeva), both in the short Abrikosov Street.

Aleksei I. Abrikosov (1875–1955) was professor of pathology. He was the director of the Institute of Morphology. He and his pathologist wife, Fanni D. Abrikosova, had one son, Aleksei A. Abrikosov (1928–2017), who shared the Nobel Prize in Physics in 2003. The physicist Abrikosov told us the story his father had told him: "One day [in 1951] I [A. I. Abrikosov] was called to the regional party committee. They asked me whether I knew that all nationalities in the Soviet Union were equal. Of course, I did. Then they told me that half of the co-workers in my Institute were Jewish, whereas in the whole population the Jews amounted to a few percent only. They told me that this situation was unjust to the other nationalities and that I should correct the situation."[2] The older Abrikosov found no excuse to fire any of his associates; they were good because he had hired them for their excellence. This was at the time of the anti-Semitic campaign during the last years of Stalin's reign, and Abrikosov resigned.

Boris V. Petrovsky (1908–2004), a physician and surgeon, was the minister of health of the Soviet Union for fifteen years (1965–1980), and was also recognized as a great surgeon of the thorax. Petrovsky's bust and the statue on his grave present his dynamic personality donning a surgeon's garb (Chapter 7). He was the director of an institute of clinical and experimental surgery in Moscow even during his tenure as health minister. In the years 1949–1951, he was head of the Third Clinic of Surgery of the University of

[2] Balazs Hargittai and Istvan Hargittai, *Candid Science V: Conversations with Famous Scientists* (London: Imperial College Press, 2005), pp. 187–188.

Budapest. One might think that his colleagues would have resented his appointment as an imposition, but he made himself popular and accepted.

Nikolai Sklifosovsky's memorial (Salavat A. Shcherbakov, 2018) at the intersection of Abrikosov and Bolshaya Pirogovskaya Streets.

In fall 2018 a memorial was erected for Nikolai V. Sklifosovsky (1836–1904). He was a Ukrainian surgeon, born in Moldova, who received his education at Moscow University. He was known for his skills in military surgery. He was director of the Imperial Clinical Institute in St. Petersburg. He helped establishing the Device Pole campus and advocated the unity of medical training and clinical practice.

Clinic of Dermatology

Memorial plaque of Viktor Rakhmanov on the façade of the University Clinical Hospital No 2 at 4 Bolshaya Pirogovskaya Street.

The Clinic of Dermatology was established in 1895. The memorial plaque on the façade honors Viktor A. Rakhmanov (1901–1969), specialist of dermatology and venereal diseases and head of department. He studied here and was professor until the end of his life.

Nikolai Semashko

Bust of Nikolai Semashko (L. V. Tazba, 1982) at 2 Bolshaya Pirogovskay Street, in front of the administrative section of the clinical center.

Nikolai A. Semashko (1874–1949) was a physician and an organizer of the health service of the Soviet Union. As a medical student of Moscow University he participated in an illegal revolutionary movement and was exiled. He graduated as medical doctor from Kazan University in 1901. Subsequently, he spent years in exile in Switzerland and in France. After the 1917 revolutions, he held high offices in the health system and was appointed the equivalent of health minister of the Russian Federation in 1918. He served in this post until 1930. Afterwards, he helped organizing new medical departments, founding a national medical library, establishing institutions for the healthy nutrition of children, and enhancing the general level of hygiene and the welfare of scientists.

Semashko's career included some negative features. In the early 1920s, he argued for the forced centralization of the medical service. In 1947, Professor Grigory Roskin and his wife, Professor Nina Klyucheva, were investigated for their role in making their new book available in manuscript to Americans. The book was about a new anti-cancer treatment. The manuscript was given to the

Americans by academician Vasily V. Parin who acted, ostensibly, with the permission of the then health minister, but Parin was severely punished for his "crime." The case, under Stalin's personal management, was a prelude to the forthcoming anti-Semitic campaigns. Semashko was one of the judges in a humiliating public investigation of Roskin and Klyucheva and was an eager participant in the case. Parin was incarcerated and Roskin and Klyucheva were vilified. Fortunately, all three survived and eventually continued in their professions.

Ivan Sechenov

Left: The Museum of the History of Medicine and the statue of Ivan Sechenov (L. E. Kerbel, 1958) in its front at 2 Bolshaya Pirogovskaya Street. I. P. Zalessky and K. M. Bykovsky designed the building in 1896 for an outpatient clinic.

Ivan M. Sechenov (1829–1905) graduated from an engineering high school and following two years of military service he enrolled at the medical school of Moscow University. He attended courses that went much beyond a physician's training and acquired a broad-based education. In 1856, after graduation, he continued his studies in Western Europe with such giants of science as Hermann von Helmholtz. In 1860, in St. Petersburg, he defended his dissertation for the Doctor of Medicine degree in the Academy of Medical Surgery. In 1870 he left the Academy and joined Dmitry Mendeleev for a short while. Between 1871 and 1876 he was head of physiology at the Novorossiisky University in Odessa. Then, between 1876 and 1888 he was professor of physiology at the University in St. Petersburg. He was an early supporter of the Higher Courses for Women.

In addition to his medical activities, Sechenov was involved in literature and contributed to the development of the Russian literary language. From 1891 he was professor of physiology at Moscow University. Kliment Timiryazev found Sechenov a towering figure among his peers with the breadth of research extending from the purely physical in determining the solubility of gases to neurophysiology and psychology. According to Timiryazev, Sechenov expressed himself simply and unambiguously, yet beautifully, and had a far-reaching impact on Russian science and thought.

The Museum of the History of Medicine was founded in 1990 and is located in a former ambulatory building. It has about 70,000 exhibits, including portraits and busts of renowned professors and physicians; medical instruments of previous centuries; anatomical atlases; medical books; and others. This Museum is unique in Moscow.

There is a small specific museum devoted to the history of psychiatry in Russia and the Soviet Union and to the history of the N. A. Alekseev Psychiatric Clinical Hospital, located at 2 Zagorodnoe Highway. This hospital was opened in 1894 at the initiative of an elected city official, Nikolai A. Alekseev. In 1922, the hospital was named after Petr P. Kashchenko (1859–1920), a psychiatrist who was in charge of psychiatric institutions from 1889, and, among them, he headed this psychiatric hospital between 1904 and 1907. On the occasion of the centennial of the hospital, it was renamed after its initiator, Nikolai A. Alekseev.

Once psychiatry in the Soviet Union is mentioned, we must remember its dark side in the last decades of Soviet reign. So-called dissidents, people fighting for the protection of human rights and for democratic change, were often declared mentally ill by the center of forensic psychiatry. They were sent for forced "treatment" in special hospitals of the interior ministry. One of the victims of such "treatment" compared it to the inhuman experiments by Nazi doctors in the German concentration camps.[3]

There are medical institutions and memorials in the Pogodinskaya Street bordering the Sechenov campus from the west. One example is the Gamaleya bust shown further down in this chapter and another is the Orekhovich memorial plaque.

Memorial plaque of Vasily N. Orekhovich at 10 Pogodinskaya Street (courtesy of Olga Dorofeeva).

[3] https://en.wikipedia.org/wiki/Serbsky_Center; accessed 01/20/17.

Vasily N. Orekhovich (1905–1997) was a biochemist who specialized in protein research. He was the long-time director of what is today the Orekhovich Research Institute of Biomedicinal Chemistry. His memorial plaque is on the façade of the Institute.

Heroes of Medical Troops

Left: Memorial of the medical troops helping the wounded in WWII (L. E. Kerbel, 1972), in the park at 2 Bolshaya Pirogovskaya Street (courtesy of Vasily Ptushenko). The cross-shaped memorial of red granite refers to the Red Cross. *Right*: Memorial plaque of the physician Nadezhda V. Troyan (1921–2011) on the façade of the Museum of the History of Medicine. She served as a medical sister for the partisans and completed her studies after the war.

Nikolai Pirogov

Left: Statue of Nikolai I. Pirogov (V. O. Shervud, 1897) in front of the University Clinical Hospital No 2 and the Fronshtein Clinic of Urology at 2 Bolshaya Pirogovskaya Street. *Right*: His bust (not far from the statue) at 1 Malaya Pirogovskaya Street.

Nikolai I. Pirogov (1810–1881), surgeon and anatomist, founded military field-surgery and anesthesiology in Russia. He studied medicine at the University of Yuryev (Yuryev was also known by its German name, Dorpat; today, it is Tartu, Estonia). Following graduation, he took advanced training in Western Europe. At 26, he was already a professor at his Alma Mater. Soon, he was invited to St. Petersburg to be head of surgery at the Academy of Military Surgery. He trained military field surgeons and worked out new approaches to operations that are carried out on the battlefield.

Pirogov developed the so-called topographical anatomy, which pays increased attention to the relationship among various components of the body, such as muscles, nerves, arteries, and others. In 1847, he joined the Russian Army and worked in a real military environment. He pioneered conducting surgery under anesthesia using ether and he carried out thousands of operations. In the Crimean war, 1853–1855, Pirogov was in charge of surgery in Sevastopol besieged by the British and French armies. Pirogov saved many fighters from amputation of their limbs. He found time for educating medical sisters. Even though his responsibilities increased, he was still active in later wars. According to the inscription on his statue, even if our soul desires to become a thorough specialist, it cannot happen without a broad humanistic education.

Grigory Zakharkin

Memorial plaque of Grigory A. Zakharin on the façade of the Fronshtein Clinic of Urology, 2 Bolshaya Pirogovskaya Street. He worked here between 1890 and 1896.

Grigory A. Zakharin (1829–1898) graduated from Moscow University; and between 1856 and 1859, he took advanced training in internal medicine in Western Europe. In 1984, he tended Alexander II for which his "leftist" peers criticized him. He was an unusual doctor who relied more on lengthy conversations with his patients than laboratory data, but his patients trusted his judgment. Such celebrities were among his patients as the writers Anton P. Chekhov — himself a physician by training — and Lev N. Tolstoy. Zakharin advanced the development of pediatrics, gynecology, and neuropathology. According to contemporary reports, he was unpleasant and uncommunicative and charged high fees for his services, but was also charitable and generous, and helped the needy without expecting gratitude. His marble bust (A. S. Golubkina, 1990) stands in the Museum of the History of Medicine.

Fedor Erisman

Bust of Fedor F. Erisman (N. S. Shevnukov, 1937) in the park at 2 Bolshaya Pirogovskaya Street (courtesy of Vasily Ptushenko).

Fedor F. (Friedrich H.) Erisman (1842–1915) was a Swiss-Russian physician who initiated hygienic methods in medicine in Russia. He married the first Russian woman physician, the gynecologist Nadezhda P. Suslova, but the marriage ended in divorce. His second marriage was happy. He was interested in the illnesses of the eye, investigated the impact of school on myopia of children, and created a new kind of school bench. He devoted his life to the science of hygiene and carried out a large volume of teaching at the medical faculty of Moscow University. The Institute of Hygiene grew out of his activities; today, it is the Erisman Research Center of Hygiene in Mytishchi, Moscow Region. In 1891, Erisman was fired from the University for his backing the student movement. He returned to Zurich where, from 1901, he was in charge of the sanitation section of the city government.

Vladimir Snegirev

Statue of Vladimir F. Snegirev (S. T. Konenkov, 1967) in front of the Snegirev Clinic of Obstetrics and Gynecology.

Vladimir F. Snegirev (1847–1917) was an orphan and could study only due to a rich patron. Snegirev graduated from Moscow University and became a professor there. He was one of the founders of surgical gynecology in Russia. He supported women becoming physicians and not only in obstetrics-gynecology. He initiated an institute and a clinic for gynecology. It opened in 1889, and it is today the Snegirev Clinic of Obstetrics and Gynecology at 2 Elansky Street. The street was named in 1965 after the surgeon Nikolai N. Elansky (1894–1964) who made important contributions to the development of field surgery during WWII. Elansky has a memorial tablet on the façade of the Snegirev Clinic.

Snegirev was an innovator in his field; he performed two thousand operations, and a number of procedures carry his name. He was popular among his patients; they adored him; he radiated his love of his profession. This being the case, it is easy to imagine how it hurt him when the family of a patient he lost carved onto her tombstone (at the Donskoe Cemetery in Moscow) that she died from an operation by Dr. Snegirev. The operation was of high risk, but it was unavoidable and the tragedy tormented Snegirev to the end of his life.

Nil Filatov

Statue of Nil F. Filatov (V. E. Tsigal, 1960) in the park at the intersection of Bolshaya Pirogovskaya and Elansky Streets and his bust (V. G. Kudrov, 1989; courtesy of Vasily Ptushenko) in front of the Filatov Municipal Clinical Hospital.

Nil F. Filatov (1847–1902) was the founder of pediatrics in Russia. He graduated from the medical school of Moscow University in 1869 and was awarded the Doctor of Medicine degree in 1876. In the period 1872–1874, he continued his education in Western Europe. From 1875 to the end of his life, he divided his time between practicing bedside medicine in pediatrics and teaching medicine at Moscow University. The first pediatric hospital in Russia started its operations in St. Petersburg in 1834 and one in Moscow followed in 1842; this was the Sofiya Pediatric Hospital. After it had been destroyed by fire, the new hospital had moved to its present location, at 15 Sadovaya-Kudrinskaya Street. In 1922, it was renamed after Filatov.

Timofei Krasnobaev

Bust of Timofei P. Krasnobaev on the campus of the Morozov Pediatric Hospital in front of the divisions of blood transfusion and oncology & hematology (courtesy of Olga Dorofeeva).

Timofei P. Krasnobaev (1865–1952) was a specialist of osteotuberculosis. He graduated from Moscow University in 1888. From 1919, he was a consultant to the Institute of Tuberculosis until the end of his life. From 1939, he was in charge of the division of surgery of the First Moscow Pediatric Hospital (today, Morozov Pediatric Hospital), 1 Dobrinsky Street. He was one of the founders of pediatric surgery in Russia.

Evgeny Tareev

Bus-relief of Evgeny Tareev on the façade of the Tareev Clinic of Nephrology, Internal, and Occupational Medicine, 11 Rossolimo Street.

Evgeny M. Tareev (1895–1986) was a specialist in internal medicine, one of the founders of nephrology, hepatology, rheumatology, and parasitology in the Soviet Union. He was professor at the Sechenov Medical University, worked also as head of a research group at the Academy of Sciences, and was one of Stalin's personal physicians.

Sergei S. Korsakov

Bust of Sergei Korsakov (S. D. Merkurov 1949) in front of the Korsakov Clinic of Nervous Diseases, 11 Rossolimo Street (courtesy of Olga Dorofeeva).

Sergei S. Korsakov (1854–1900) graduated from Moscow University in 1875 and continued his career under his mentor, A. Ya. Kozhevnikov. Korsakov's field was mental diseases and he traveled to France and Germany to keep abreast with new developments. He applied the English approach of a no restraint regime to his patients and further developed it. He introduced a regime without forced limitations, without closed doors, and avoided the use of the strait-jacket. He trained future psychiatrists and neuro-pathologists. His view was that a mentally ill person should not be considered just a number; rather, such a patient should be treated as an individual and should be well known by all who have any part in the treatment of that patient.

Nikolai Gamaleya

Adjacent to the campus of the Sechenov Medical University is the Polyclinic of the Gertsen Oncological Institute at 6 Pogodinskaya Street. In its front, in a little garden, there is a bust of the epidemiologist and microbiologist Nikolai F. Gamaleya (1859–1949).

Left: Bust of Nikolai F. Gamaleya (S. Ya. Kovner and N. A. Maksimchenko, 1956) in front of the Polyclinic of the Gertsen Oncological Institute. *Right*: Bust of Ilya I. Mechnikov, at the Institut Pasteur in Paris. The bust was a gift of the Ukrainian Academy of Sciences.

Gamaleya graduated from Novorossiisky University in Odessa in 1880 and the Military Medical Academy in St. Petersburg in 1883. He traveled to Paris in 1885 and worked at the Institut Pasteur; he investigated rabies for a year. Upon his return to Odessa, Gamaleya, the immunologist Ilya (Élie) I. Mechnikov (1845–1916), and the bacteriologist Yakov Yu. Bardakh (1857–1929), established, with Pasteur's support, the first bacteriological station in Russia. They vaccinated people against rabies. Mechnikov at the time was back in Odessa after his various travels. In 1888 he left Russia for good and spent the rest of his life at the Institut Pasteur. In 1908, he was co-recipient of the Nobel Prize in Physiology or Medicine for his discoveries in immunology.

In the 1880s, Pasteur's teachings had not yet gained general acceptance. When hesitations led to an investigation in England, Gamaleya traveled there in 1887 to provide evidence in Pasteur's support. For the next five years, Gamaleya commuted between Paris and Odessa until the reliability of Pasteur's techniques was fully demonstrated. In 1892, Gamaleya settled in Odessa, defended his dissertation for the degree Doctor of Medicine, and between 1899 and 1908 directed the Bacteriological Institute.

From 1912, Gamaleya lived in St. Petersburg. Support for his work did not dwindle after the revolutions in 1917, and his vaccination movement against a variety of infectious diseases spread over the whole Soviet Union. Between 1930 and 1939 he directed what is today the Gamaleya Research Center of Epidemiology and Microbiology, 8 Gamaleya Street. The predecessor of this Center started its operations in 1891 and many of the world-renowned Russian and Soviet microbiologists and epidemiologists worked in it. Gamaleya was also a professor of microbiology at the Second Moscow Medical Institute.

Gamaleya Street

There is a section called "Medgamal" of the Gamaleya Research Center of Epidemiology and Microbiology in Gamaleya Street. This section is one of the oldest institutions in Russia involved in the production of immunological products. In 1937 the Soviet government decided to create an independent organization for the production of bacterial products. It became part of the Gamaleya Research Center in 1952. We mention here five memorial plaques. All are shown below, three are discussed briefly in addition to the more detailed discussion of Gamaleya above, followed by another more detailed discussion of Lev A. Zilber.

On the façade of "Medgamal," *from left to right*: Gamaleya; E. N. Pavlovsky; S. V. Prozorovsky; and V. D. Timakov (courtesy of Olga Dorofeeva).

Evgeny N. Pavlovsky (1884–1965) was a zoologist specializing in entomology. He graduated from the Military Medical Academy in St. Petersburg and acquired a high rank in the medical service of the armed forces. His research focused on the epidemics caused by parasites and transmitted diseases in Central Asia, Transcaucasia, the Crimea, and the Far East. He directed the Institute of Zoology in Leningrad, was the president of the

Soviet Geographical Society, and the president of the Soviet Entomological Society.

Sergei V. Prozorovsky (1931–1997) was a microbiologist who worked at the Gamaleya Center from 1958 and was its director from 1983 until his death.

Vladimir D. Timakov (1905–1977) created a school of microbiologists and epidemiologists. Between 1968 and 1977 he was president of the Academy of Medical Sciences. During WWII he was minister of health of Turkmenistan. After the war, he was appointed director of the Gamaleya Center of Epidemiology and Microbiology. There was considerable improvement in the production of medicines and prophylactic products under his stewardship. From 1949, he was also professor and head of the department of microbiology at the Second Moscow Medical Institute.

Zilber's memorial plaque on the façade of "Medgamal" (courtesy of Olga Dorofeeva).

Lev A. Zilber (1894–1966) graduated both in medicine and in natural science from Moscow University. In 1928, he and his scientist wife, Zinaida V. Ermoleva (Chapter 8), visited the Pasteur Institute in France and the Koch Institute in Germany. The marriage soon ended in divorce. Zilber became head of microbiology in Baku. He was first arrested in 1930, falsely accused with infecting the people of Azerbaijan with a plague, whose epidemic he had in reality been trying to suppress and for which he had received an important decoration. He was freed after four months of incarceration.

He moved to Moscow and created a division of virology at the Institute of Microbiology of the Academy of Sciences. In 1935, he married again; Valeriya P. Kiseleva became his second wife. In 1937 Zilber went for an expedition to the Far East to study an epidemic, which turned out to be tick-borne encephalitis — a human viral infectious disease of the central nervous system. It was caused by a heretofore unknown virus, identified for the first time in 1937. One of its three known sub-types is the Far eastern tick-borne encephalitis virus (formerly known as the Russian Spring Summer encephalitis virus).

Upon his return to Moscow, Zilber was arrested again. This time the principal accusation was that he allegedly wanted to infect the population of the capital city by introducing encephalitis into the water mains of Moscow. He was sent to a slave labor camp where hundreds of inmates were dying from vitamin deficiency. Zilber saved them by his rudimentary concoction prepared from ingredients available on the spot, for which he then received an author's certificate (the certificate was issued formally for the security organs). In 1939, he was freed and appointed head of the virology section of the Central Institute of Epidemiology and Microbiology of the Ministry of Health.

In 1940, Zilber was arrested for the third time, for declining work on bacteriological weapons. He was sent to a *sharashka*, a slave labor camp where the prisoners were engaged in meaningful occupation, and he did cancer research. He concluded that certain cancers are initiated by viral infection, which was a revolutionary discovery at the time. In the meantime, there was the German aggression in June 1941 and the battle of Moscow in fall 1941. The retreating Germans deported Zilber's family from a suburb of Moscow and the family spent the next three and half years in labor camps. Zilber was freed in 1944. Several influential personalities acted on his behalf, among them, Nikolai Burdenko, the chief surgeon of the Red Army; Leon Oberli, one of the vice presidents of the Academy of Sciences; the biochemist Vladimir Engelhardt; and Zilber's ex-wife, Zinaida Ermoleva who had played a principal role in developing the production of penicillin in the Soviet Union. Zilber embarked on finding and bringing back his family and he succeeded miraculously.

In 1945, Zilber was elected member of the newly organized Academy of Medical Sciences and appointed head of its Institute of Virology and head of the section of tumor virology and immunology of the Institute of Epidemiology, Microbiology, and Infectious Diseases. However, during the post-war period until Stalin's death in 1953, Zilber could never be sure if and when he might be arrested with the most arbitrary accusations. This did not happen, but his creativity suffered. His last fourteen years, from 1953 until his death were the most peaceful period of Zilber's life. Both of Zilber's sons, Lev L. Kiselev and Fedor L. Kiselev became renowned molecular biologists (Chapter 1).

Other Memorials

We start with the memorials of Vladimir Demikhov. The memorial tablet on the church at 6 Bolshaya Pirogovskaya Street (see below) bridges nicely the memorials in the Deviche Pole campus and those found elsewhere in Moscow.

Vladimir Demikhov

Statue of Vladimir Demikhov (D. A. Stritovich, 2016) at the Shumakov Institute of Transplantology and Artificial Organs (courtesy of Darya Kobyatskaya).

Vladimir P. Demikhov (1916–1998) was a biologist and surgeon who pioneered organ transplantation. His father fell in WWI when Demikhov was three years old. He studied physiology at the biology faculty of Moscow University. As a third-year student, he already constructed an artificial heart and implanted it into a dog that survived with it for two hours. Demikhov graduated in 1940 and started his research, but when the war came, he served on the front.

After the war he worked at the Institute of Experimental and Clinical Surgery. He transplanted hearts in dogs, which became a world sensation though he received little attention in the Soviet Union. Between 1955 and 1960 he worked at the First Moscow Medical Institute whose director did not let Demikhov defend his PhD-equivalent dissertation about his experiments in organ transplantation. Demikhov was too much of an innovator and by training he was a biologist and not a physician. He had to move to the Sklifosovsky Institute of Emergency Services, where he stayed until 1986. There is now a memorial plaque there honoring the "prominent scientist-transplantologist" (see, below).

Demikhov published the essence of his dissertational work in a book, *Experimental transplantation of vital organs*, in 1960 in Russian and its English translation appeared in the West in 1962. Demidov coined the word transplantology. His international recognition was much more significant than in his own country. *TIME* magazine wrote about him already in 1955.

He was finally allowed to defend his dissertation in 1963 though not in a medicinal institution but at the Faculty of Biology of Moscow University. He was granted at once the higher degree of DSc in the biological sciences.

The persecution he had been suffering for his unusual scientific activities took its toll, and he suffered a stroke in 1968. His laboratory continued its activities until 1986 in organ transplantation. His international recognitions included honorary membership in the Royal Uppsala Scientific Society in Sweden and the famous American Mayo Clinic, among others. Christiaan Barnard who in 1967 performed the first human-to-human heart transplant operation visited Demikhov's laboratory in 1960 and 1963, not officially, but as a tourist. He even assisted Demikhov in an operation in order to gain experience. Barnard has been quoted saying, "if there is a father of heart and lung transplantation then Demikhov certainly deserves it."[4] Demikhov died in 1998 and his grave is in the Vagankovskoe Cemetery (section 24). If he could not be recognized in his life by the medical profession of his country, at least his tombstone says, "surgeon." In 2016, his statue was unveiled on the seventh floor in a new pavilion of the Shumakov Institute of Transplantology and Artificial Organs, Shchukinskaya Street 1.

Top: The Holy Dmitry Prilutsky Eastern Orthodox Church. *Bottom*: Memorial tablet on the façade of the church honors Vladimir Demikhov.

[4] See, T. A. Fricke and I. E. Konstantinov, In Marco Picichè, Ed., *Dawn and Evolution of Cardiac Procedures: Research Avenues in Cardiac Surgery and Interventional Cardiology* (Springer, 2012), p. 82.

Lately, there is increasing recognition of Demikhov's achievements. Even Russia recognized him shortly before he died with a high state award. The Eastern Orthodox Church erected a memorial tablet on the facade of the church at 6 Bolshaya Pirogovskaya Street. This building of Russian-Byzantine style was built in about 1880 and consecrated as a church in 1903. It was designated as the chapel to serve the students and instructors of Moscow University. During the Soviet era the building was part of the First Moscow Medical Institute. Following the dissolution of the Soviet Union, the building was returned to the Eastern Orthodox Church.

This was the building where Demikhov was running his laboratory while he was at the First Moscow Medical Institute. The memorial tablet says that Demikhov's laboratory operated in this venue in the 1950s and 1960s and it was at the time when the Church was being persecuted. From the magnanimous inscription one perceives pride as it describes that "here is where the great Russian scientist, the founder of transplantology in the world, Vladimir Petrovich Demikhov (1916–1998), conducted the world's first successful experiments of transplanting organs. May this promoter of science be remembered forever."

Sklifosovsky Institute

Central section of the initial building of the Sklifosovsky Research Institute of Emergency Services, courtesy of Olga Dorofeeva.

The Sklifosovsky Research Institute of Emergency Services at 3 Bolshaya Sukharevskaya Square was originally a mansion of the aristocrat Nikolai Sheremetevo. The name of the building was Strannopriimny Dom and the inscription is still visible on the rotunda. It was meant to be a free-of-charge hospital for the poor and opened in 1810 at the initiative of the wife of the aristocrat Praskov Zhemchugov. On the basis of this institution, the Sklifosovsky Institute was established in 1923.

Memorial plaques of Sergei S. Yudin (*left*) and Boris A. Petrov (*right*) at the Sklifosovsky Institute; courtesy of Olga Dorofeeva.

Sergei S. Yudin (1891–1954) was a surgeon, director of the Vishnevsky Institute of Surgery and one of the leading surgeons of the Sklifosovsky Institute specializing in stomach surgery. In WWII, he was a senior consultant to the chief surgeon of the Red Army. In addition to domestic recognition, he was an honorary member of the Royal College of Surgeons (UK), the American Association of Surgeons, the Society of Surgeons at the University of Paris, and other learned societies. In 1946, the Sorbonne honored him with an honorary doctorate.

In 1948, Yudin was arrested, accused of espionage for Britain, and sentenced to death. He was incarcerated in the Lubyanka, then moved to the Lefortovo prison where he was kept in solitary confinement and where he suffered a myocardial infarction. Yudin's death sentence was commuted to exile after he had agreed to declare himself an anti-Semite and to accuse a well-known military physician of Jewish nationalism. In exile, Yudin worked as a surgeon in Siberia. He was freed after Stalin's death, but died shortly after.

Boris A. Petrov (1898–1973) was the chief surgeon of the Soviet Navy. He pioneered radical surgery in the treatment of patients suffering from stomach cancer. He established the practice of artificial replacement of excised parts of the stomach by plastic surgery.

Memorial plaque of Vladimir Demikhov at the Sklifosovsky Institute (courtesy of Olga Dorofeeva).

We have seen that at some point, the pioneer transplantologist Vladimir Demikhov, had to leave the First Moscow Medical Institute and he moved to the Sklifosovsky Institute where he continued his research until the end of his professional life. His memorial tablet at the Institute was erected by private funds provided by Professor Mogeli Sh. Hubutiya (1946–), himself a noted transplantologist and a former pupil of Valery I. Shumakov.

Valery Shumakov

Statue of Valery Shumakov in surgeon's garb (D. A. Stritovich, 2011) at the Shumakov Institute, 1 Shchukinskaya Street.

Valery I. Shumakov (1931–2008), future transplantologist, enrolled at the First Moscow Medical Institute in 1950 and early on started his apprenticeship with Boris Petrovsky (see above). Petrovsky's every new appointment to higher positions in the hierarchy of health institutions opened up new possibilities for Shumakov. He worked in Petrovsky's institute from 1963; he was head of the laboratory of artificial heart and supporting blood circulation, then head of the section of transplantology and artificial organs. At Petrovsky's initiative, in the late 1960s, a new institute was organized for transplantology and artificial organs and in 1974 Shumakov moved there and became its director. He had several other functions and positions, among them, he was head of the Department of Living Systems at the Moscow Institute of Physics and Technology. During the last years of his life, Shumakov's home was in the House on the Embankment (Chapter 8).

Nikolai Burdenko and His Institutions

Principal building of the Burdenko Military Hospital at 3 Gospitalnaya Square and its Rotunda with the bust of Peter I (courtesy of Marina Ovchinnikova).

The predecessor of the Burdenko Main Military Clinical Hospital was the first state institution in Russia to treat the sick and to train Russian physicians in a structured setting. The first Russian professors of medicine who used their mother tongue in practicing their profession appeared here. On May 25, 1706, Peter I ordered the establishment of this institution, which accepted its first patients on November 21, 1707. Next to the first makeshift ward, there was a chapel and a botanical garden for cultivating the plants from which the ingredients for medicines were extracted.

Bust of Peter I and the memorial of Peter I and Nikolai Bidloo (L. Baranov, 2008) in the park of the Burdenko Military Hospital (courtesy of Marina Ovchinnikova).

Nikolai (Nicolaas) Bidloo (1669 or 1670–1735) was a Dutch medical doctor and the court physician of Peter I. For 30 years, he was in charge of the first Russian medical establishment. He was born in Amsterdam into a medical family and in 1702 he signed a contract to move to Moscow as Peter's personal physician. Beside directing the predecessor of today's Burdenko Hospital, Bidloo initiated a school of clinical medicine for 50 students. He taught anatomy and surgery, and assembled the best possible student body even if it took raiding other institutions for their most talented pupils.

Left: Memorial of military physicians on the occasion of the 200-year anniversary of the 1812 Patriotic War (courtesy of Olga Dorofeeva). *Right*: Statue of Nikolai I. Pirogov in a surgeon's garb (courtesy of Marina Ovchinnikova). Both memorials are at the Burdenko Military Hospital.

Busts of Nikolai N. Burdenko in the park of the Burdenko Military Hospital (*left*, courtesy of Marina Ovchinnikova) and in the garden of the Burdenko Institute of Neurosurgery (*right*, S. D. Merkurov, 1949).

Nikolai N. Burdenko (1876–1946) studied in a divinity school, but for higher education, he opted for medical school. He attended Tomsk University, but was expelled for participating in a radical student movement. Finally, he graduated as a physician in 1906 at the University of Yuryev (today, Tartu, Estonia). By 1917, he was back there as a professor, but not for long as Estonia was to become independent and he and his clinic moved to Voronezh in Russia proper.

Burdenko was interested in military medicine and in the preparedness of the medical personnel of the Red Army. He relied on his experience in the Russo-Japanese war and in WWI. He moved from Voronezh to Moscow in 1923 and the next year he was appointed head of surgery at what is today the Burdenko Clinic of Surgery of the Sechenov Medical University, which is yet another institution that carries Burdenko's name. He kept this position to the end of his life. In 1929, he was appointed to be director of what is today the Burdenko Institute of Neurosurgery. He initiated and supervised the establishment of a network of neurosurgery clinics in the Soviet Union. In WWII, he first served on the Finnish front and then treated wounded soldiers returning from the German front. He suffered a stroke, but made efforts to carry on. He initiated the establishment of the Academy of Medical Sciences in 1944 and was its first president.

A dark stain on Burdenko's career was his chairmanship of the Extraordinary State Commission, referred to often as the "Burdenko Commission." It was charged in 1944 with investigating the killings of twenty-two thousand Polish officers, doctors, professors, lawmakers, police officers, and other members of the Polish elite in 1940 in the Katyn Forest. The Commission "found" overwhelming evidence that the Germans committed the crime. However, over the ensuing decades doubts and evidences kept surfacing pointing to falsification by the Burdenko Commission of what happened, because it was the Soviet security organs that committed the murders. Mikhail Gorbachev admitted Soviet responsibility, but would not accuse Stalin directly with complicity. Under Boris Yeltsin, the execution order by Stalin was declassified, but in the ensuing years, the policy of stressing Soviet achievements rather than uncovering the atrocities prevailed. Finally, in 2010, seventy years after the events and as a result of extended investigations, the State Duma — the Russian Parliament — accepted a resolution. It established that the Katyn massacre was carried out on Stalin's direct orders and with the direct involvement of other top Soviet leaders.

Busts of high-ranking military physicians in the park of the Burdenko Military Hospital, *from left to right*: Miron S. Vovsi, Efim I. Smirnov, and Ivan A. Yurov (courtesy of Olga Dorofeeva).

Miron S. Vovsi (1897–1960) studied medicine first at the University of Yuryev (today, Tartu), and graduated as a physician from Moscow University in 1919. He was in the Red Army right from his graduation, eventually rising to the rank of general. In the years 1941–1950, he was the main physician of internal medicine of the Red Army. Then, he worked as consultant in the Kremlin Hospital. He was arrested shortly before Stalin's death in connection with the so-called "doctors' plot," was subjected to torture, and declared to be head of an anti-Soviet terrorist organization. He was freed after Stalin's death and he continued his activities in the Botkin Hospital for the rest of his career. He was the model for the role of Professor Kuzmin in Mikhail A. Bulgakov's drama *Master and Margarita*.

Efim I. Smirnov (1904–1989) was a military physician and a general. Initially, for years, he was a factory worker without any professional training. In 1928, he was called to the Red Army and directed to the Military Medical Academy, from which he graduated in 1932. During WWII, he was commander of the sanitary service of the Red Army and was credited with the fact that the Army suffered no major epidemic during the war. He rose to minister of health of the Soviet Union in 1947 and his removal from this position was connected with the so-called "doctors' plot." After Stalin's death, he continued in high positions in the military health service, including the training of military physicians.

Ivan A. Yurov (1921–1986) was a general of the medical service of the Soviet Army and an organizer of the Soviet Military Medical Service.

The initial Burdenko Institute of Neurosurgery at 16 Fourth Tverskaya-Yamskaya Street.

Three of the memorial plaques on the façade: *from left to right*: Boris G. Egorov, Aleksandr I. Arutyunov, and Fedor A. Serbinenko.

What is today the Burdenko Institute of Neurosurgery was begun in 1932. It was a research center not only of neurosurgery, but also of neuroanatomy and psychiatry. The memorial plaques on the façade of the old building inform about physicians of high position and not necessarily about the most creative scientists. An institute museum operates on the site of the Institute.

Boris G. Egorov (1892–1972) neurosurgeon graduated from Moscow University where P. A. Gertsen and N. N. Burdenko were among his professors. Egorov was director of the Burdenko Institute between 1947 and 1964. His physician son, Boris B. Egorov, was a cosmonaut.

Aleksandr I. Arutyunov (1904–1975) neurosurgeon moved to Moscow from a provincial town in 1932. He acquired his scientific degrees at the Burdenko Institute and served as its director from 1964 for the rest of his life.

Fedor A. Serbinenko (1928–2002) specialized in interventional cardiology. His family suffered heavy losses in WWII; he had to work as a teenager, and started higher education when he was 20. He graduated in 1954 and worked at the Burdenko Institute throughout his entire career.

Ippolit Davydovsky

Bust of Ippolit V. Davydovsky (A. S. Allakhverdyants, 1974) in the hospital garden at 11 Yauzkaya Street (courtesy of Vasily Ptushenko).

Ippolit V. Davydovsky (1887–1968) graduated from Moscow University in 1910. After his medical military service in WWI, he continued his career at Moscow University, at the pathology department, headed by Aleksei Abrikosov. From 1930, Davydovsky was head of the department of pathological anatomy of the Second Moscow Medical Institute. He introduced new approaches in the utilization of pathological information from autopsies and founded a nationwide pathologist-anatomist service.

Fedor Gaaz

Bust of Fedor P. Gaaz (N. A. Andreev, 1909) at 5 Maly Kazenny Street, in front of the Research Institute of Hygiene and Protection of the Health of Children and Adolescents (courtesy of Vasily Ptushenko).

Fedor P. Gaaz (Friedrich-Joseph Haass, 1780–1853), German-born physician, attended school in Cologne, studied physics and philosophy in Jena, and medicine in Gottingen. He started his career in Vienna and at the invitation of a Russian aristocrat he moved to Russia in 1806 as a family doctor. Gaaz served as surgeon in the Russian Army in the 1812 Patriotic War. From 1813, he lived in Moscow. First he was in charge of the emergency service and from 1829 he was in charge of medical services for the Moscow prisons. He dedicated his life to improving the living conditions of the incarcerated. Towards the end of his life he extended his benevolent activities to the poor and homeless. People referred to him as the holy doctor. He is buried in the Vvedenskoe Cemetery. The stand of his bust carries his favorite saying, "Speshite delat' dobro" — Hurry to do good deeds.

Vishnevsky Institute

Statue of Aleksandr V. Vishnevsky (S. T. Konenkov, 1951) in front of the A. V. Vishnevsky Institute of Surgery at 27 Bolshaya Septukhovskaya Street (courtesy of Olga Dorofeeva).

Aleksandr V. Vishnevsky (1874–1948) was a military surgeon and the sculptor Sergei Konenkov presented him in a surgeon's garb as if Vishnevsky had just completed a tiring operation. Vishnevsky's name is also connected with an ointment he invented. It is an antiseptic, which enhances the blood flow to the wound, and its oil component facilitates the penetration of the ingredients into the depths of the skin.

Three memorial plaques of surgeons on the façade of the Vishnevsky Institute (courtesy of Olga Dorofeeva). *From left to right*: Aleksandr A. Vishnevsky (1906–1975), the son of A. V. Vishnevsky, and a general of the medical corps; Donat S. Sarkisov; and Mikhail I. Kuzin.

Donat S. Sarkisov (1924–2000) graduated from the Naval Medical Academy in 1947. He became a specialist in burn wounds and in the consequences of brain damage in connection with pneumonia. He worked at the Vishnevsky Institute of Surgery from 1957.

Mikhail I. Kuzin (1916–2009) graduated from the Military Medical Academy in Leningrad and his mentor, Nikolai Elansky, advised him to continue his education. However, the war came and as soon as Kuzin enrolled for his postgraduate studies, he turned from postgraduate student into a field surgeon. He distinguished himself in WWII and completed his education in military medicine after the war.

In the mid-1950s, Kuzin took part in the investigation of the consequences of the atmospheric nuclear explosions on September 14, 1954. The Soviet military conducted exercises in the Totskoe proving ground and field maneuvers in the Orenburg Region. This particular proving ground was selected for the test explosions because of the similarity of the terrain to the terrain of Western Europe where the Soviet military leadership supposed WWIII might begin. Marshall Georgy K. Zhukov, the supreme military leader of the Soviet Union, was in charge. There were about 45 thousand troops and ten thousand local civilians who might have been affected by these exercises. This amounted to a large-scale experiment on humans. In time, a memorial was erected on the proving ground in remembrance of those who suffered radiation damage from the test explosions. Kuzin used information from his experience in his Doctor of Science dissertation, which remained classified for a long time and some of its details may still be classified.

Kuzin received a professorial appointment at the Burdenko Clinic of Surgery at the First Moscow Medical Institute and he was rector of the Institute from 1966 until 1974. He was appointed director of the Vishnevsky Institute in 1976 and served in that position until 1988. He was also Surgeon General of the Soviet Ministry of Health.

Aleksandr Myasnikov

Bust of Aleksandr L. Myasnikov (M. P. Olenin, 1973) in front of the National Research Center for Preventive Medicine, 10 Petroverigsky Street.

Aleksandr L. Myasnikov (1899–1965) graduated from the First Moscow Medical Institute, where he eventually held a professorial appointment. Between 1938 and 1948 he worked in the medical establishment of the Soviet Navy. In 1948, he became director of the Institute of Internal Medicine of the Academy of Medical Sciences in Moscow. Today, it is the Myasnikov Institute of Cardiology at 15a Third Cherepkovskaya Street. He was a member of the physician team at Stalin's bedside during the last days of the dictator. Myasnikov completed his memoirs not long before his death in 1965, but his observations could be published only in 2011 about how a sick man was governing the Soviet empire.

Sergei Botkin and the Botkin Hospital

Left: Bust of Sergei P. Botkin (A. A. Bichukov, 1985; courtesy of Vasily Ptushenko) in front of the Botkin Municipal Clinical Hospital, 5 Second Botkinsky Street 5. *Middle*: Botkin's birth place at 35 Zemlyanoi Val Street with a memorial plaque on its façade (*right*, courtesy of Olga Dorofeeva).

Sergei P. Botkin (1832–1889) was a court physician and a specialist in internal medicine. His teachings were about the human organism as a whole over which the human will has control. He studied at Moscow University and forged a close friendship with Ivan Sechenov. In 1854, Botkin participated in stamping out the cholera epidemics. He graduated in 1855 as a physician and went to the Crimean war in Pirogov's unit (see above). Botkin held strong views on the concept of war-time medicine and the correct diet for soldiers. He perfected his medical training in Germany and France. Upon his return, he worked in St. Petersburg, and remained active not only in curing the sick but in organizational and pedagogical activities as well. He did pioneering work in promoting women's medical education and in 1876 he organized the Physicians' Courses for Women.

Today, the Municipal Clinical Hospital at 5 Second Botkinsky Street — a large medical institution with sections throughout Moscow — carries Sergei Botkin's name. The main campus has a number of memorials of which a few are presented here. The origin of the Botkin Hospital goes back to the generosity of Kozma Soldatenkov at the beginning of the twentieth century.

Statue of Kozma T. Soldatenkov on the campus of the Botkin Hospital (courtesy of Vasily Ptushenko).

Kozma T. Soldatenkov (1818–1901) was a well-to-do entrepreneur, art collector, most involved in the textile industry and book publishing. In his will, he left a substantial sum for the establishment of a hospital for people who could not afford paid treatment. The hospital was opened in 1910. Until 1920, it was called the Soldatenkovsky Hospital; today, it is the S. P. Botkin Municipal Clinical Hospital. Soldatenkov's statue was erected at its administration building in 1992.

Memorial of Fedor A. Getye on the campus of the Botkin Hospital (courtesy of Vasily Ptushenko).

Fedor A. Getye (1863–1938) physician and specialist in internal medicine was active in establishing the Soldatenkovsky Hospital. When it started its operations, Getye was appointed its physician in chief. He was a personal physician of Lenin and his family and in the 1920s he treated Lev D. Trotsky and his family. Getye was among the initiators of the establishment of the Kremlin Hospital.

Left: Memorial plaque of Anatoly P. Frumkin on the campus of the Botkin Hospital (courtesy of Vasily Ptushenko). *Right*: Bust of Aleksei D. Ochkin (Z. I. Azgur, 1955) on the campus of the Botkin Hospital (courtesy of Vasily Ptushenko).

Anatoly P. Frumkin (1897–1962) urologist graduated from Moscow University in 1921. He worked at the Botkin Hospital and made his name known for his expertise in urological surgery. During WWII he was the chief urologist of the Red Army and had the rank of colonel of the military medical service. He summarized his experience at the war front in a book. The Royal Swedish Medical Society and other learned societies elected him honorary member in recognition for his broad research and pedagogical activities.

Aleksei D. Ochkin (1886–1952) served as a physician in WWI and in the Civil War. Among his functions, he was a member of the Kremlin medical service between 1934 and 1952, being in charge of surgery and oncology during the last three years. There is another Ochkin bust by the same sculptor on Ochkin's grave at the Novodeviche Cemetery.

Aleksandr Evdokimov

Memorial of Aleksandr I. Evdokimov at 9 Vuchetich Street, in front of the stomatological school (courtesy of Olga Dorofeeva).

Aleksandr I. Evdokimov (1883–1979) first studied for medical orderly, then graduated from the Moscow Dental School in 1912 and stayed on to become an instructor. He then continued in what is now Tartu, Estonia, whose university moved to Voronezh after Estonia had gained independence. Evdokimov served as a physician in the Red Army during the Civil War. In 1922 he was appointed director of the State Institute of Stomatology and Odontology. This was the beginning of the Moscow University of Stomatology. Evdokimov worked in other medical institutions from 1930 and in 1938 he returned to the University of Stomatology and stayed there for the rest of his career.

Svaytoslav Fedorov

Statue of Svyatoslav N. Fedorov in front of a map of Russia at the Fedorov Institute of Microsurgery of the Eye (courtesy of Olga Dorofeeva).

Svyatoslav N. Fedorov (1927–2000) ophthalmologist and specialist in microsurgery started his career in Rostov, continued in Chebksary, and then in Arkhangelsk. He performed ophthalmological surgery for the first time in 1960; in 1962, he co-created a new kind of rigid lens. He moved to Moscow in 1967. By 1973, he pioneered an operation to correct early-stage glaucoma. In 1974, he created an independent laboratory for experimental and clinical surgery of the eye, which by 1979 grew into the Moscow Research Institute of Microsurgery of the Eye. In 1986 it became a scientific-technological complex on a grandiose scale in a building built specifically for his institution at 59a Beskudnikovsky Boulevard. This institution has served domestic as well as international patients. He was also head of the department of ophthalmology at the University of Stomatology (apparently this School was flexible enough to accommodate such a department). Fedorov was killed in a helicopter accident in 2000. In 2009, the number of ophthalmological surgeries following his protocol passed the five million mark. The institution founded by him is now called the Academician S. N. Fedorov Institute of Microsurgery of the Eye.

8 Leninsky Avenue and Elsewhere

There is a large medical campus at 8 Leninsky Avenue, extended around the First Municipal Hospital, which opened in 1833.

Memorial of Sergei I. Spasokukotsky (*left*, V. V. Lishev and E. F. Belashova, 1947) in front of the First City Hospital (*right*). This initial structure is awaiting renovation (Fall 2016).

Sergei I. Spasokukotsky (1870–1943) was a surgeon, a specialist in pulmonary and gastroenteric operations. He created a clinical school of his pupils and followers. He was concerned with blood transfusion and the application of his teachings saved thousands of lives in WWII. He and his fellow surgeon I. G.

Kochergin proposed a fast method of disinfecting the surgeon's hands, which proved especially useful in war-time field operations. Spasokukotsky was head of a department of surgery at the Second Moscow Medical Institute, which now bears his name.

Bust of Mikhail I. Averbakh (S. D. Merkurov, 1952; courtesy of Olga Dorofeeva) at the Helmholtz Research Institute. His memorial tablet is on the façade of the Averbakh Clinic of Ophthalmology at 8 Leninsky Avenue.

Mikhail I. Averbakh (1872–1944) ophthalmologist graduated from Moscow University in 1895. He began his career in the V. A. and A. A. Alekseevskaya Eye Hospital[5] with which he remained connected throughout his eventful career. In 1910, Averbakh founded in this hospital a department of eye diseases for the Higher Courses for Women — it then became part of the Second Moscow Medical Institute. He held also a professorial appointment at Moscow University and was among the professors who in 1911 left the University in protest against the actions of the minister of education curtailing university autonomy.

When, at Averbakh's initiative, the Alekseevskaya Hospital was transformed into the Helmholtz Research Institute, he became its first director. The Institute was charged with the nationwide coordination of the work of all ophthalmological institutions. Averbakh was active in a broad spectrum of eye diseases, such as glaucoma and trachoma, and was concerned with the treatment of eye injuries and blindness. He was Lenin's ophthalmologist and treated him repeatedly.

[5]The original building of the hospital still stands, recently renovated, at the corner of Furmanny and Sadovaya-Chernogryazskaya Streets, and is now part of the Helmholtz Research Institute of Eye Diseases.

Bust of Lyudvig I. Sverzhevsky (A. N. Burganov, 2011; courtesy of Olga Dorofeeva) at the Sverzhensky Research and Clinical Institute of Otolaryngology. His memorial tablet is on the façade of the Sverzhevsky Clinic of the Diseases of the Ear, Throat, and Nose at 8 Leninsky Avenue.

Lyudvig I. Sverzhensky (1867–1941) was an otolaryngologist who graduated from Moscow University in 1893 and continued his studies in international venues. He received professorial appointment at Moscow University in 1903, but in 1911 he moved to the Higher Courses for Women where he became head of the department of otolaryngology. Sverzhensky introduced innovations and developed a school of otolaryngology that had an impact on this area of medicine for the whole of the country. Today, there is the Sverzhensky Research and Clinical Institute of Otolaryngology at 18a Zagorodnoe Highway.

Memorial plaque of Boris S. Preobrazhensky on the façade of the First Municipal Hospital.

Boris S. Preobrazhensky (1892–1970) otolaryngologist graduated from Moscow University in 1914. Eventually, he was head of the department of the diseases of the ear, throat, and nose at the Second Moscow Medical Institute. For twenty years Stalin was among his patients and in 1948 Preobrazhensky was appointed chief otolaryngologist of the Kremlin Hospital. During WWII, he directed the otolaryngologist services of evacuated hospitals nationally. In 1952–1953 he was persecuted in the so-called "doctors' plot" and in 1953 he was arrested for the alleged mistreatment of Soviet leaders. Immediately after Stalin's death, Preobrazhensky was freed. He was an internationally recognized expert in his field, introduced a number of innovations in surgery, and created a school of specialists in otolaryngology.

Left: Bust of Aleksandr Bakulev (G. I. Ozolina, 1976) in front of the Bakulev Research Institute of Cardiovascular Surgery (courtesy of Olga Dorofeeva). *Right*: Bakulev's memorial plaque (Z. Tsereteli, 2016) at 1 Kudrinskaya Square. (https://dkn.mos.ru/presscenter/news/detail/3154026.html; Creative Commons; accessed 02/04/17).

Aleksandr N. Bakulev (1890–1967) graduated from the medical school of Saratov University. From 1919, he worked under Sergei Spasokukotsky, then from 1926 in the department of surgery of the Second Moscow Medical Institute. He became head of this department in 1943. He developed a school of surgeons in the Soviet Union and for the first time in the country he performed operations to correct congenital disorders of the heart. He was a versatile specialist who conducted surgery for the nervous system, pulmonology, heart problems, and others.

Veterinary Science

This section could also be part of Chapter 6 of agricultural science. However, Chapter 6 is on the memorials found entirely at the Timiryazev Academy whereas the present chapter extends over many areas of Moscow, hence this section was felt less out of place here. Furthermore, the best-known story mentioned here relates to humans just as much as to animals, and this provided an additional reason to include this section in this chapter.

What is today the Kovalenko Research Institute of Experimental Veterinary Science at 24 Ryazansky Avenue, was founded in 1898 in St. Petersburg, and moved to Moscow in 1918. From 1981, it carries the name of Yakov R. Kovalenko. The main purpose of the Institute has been the development of the theoretical foundations and methodology of fighting dangerous infectious illnesses of animals with emphasis on the needs of the food industry. The old building of the Institute in Kuzminsky Park is no longer in working order, but at the time of collecting images of the memorial plaques (Summer 2017), the plaques were still on the façade of this abandoned building.

Memorial plaques, from *left to right*, of Ilya I. Ivanov, Sergei N. Vyshelessky, Konstantin I. Skryabin, and Yakov R. Kovalenko (courtesy of Olga Dorofeeva).

Ilya I. Ivanov (1870–1932) was a biologist who specialized in artificial insemination and in the hybridization of humans and apes, a most controversial project. He graduated from Kharkov University in 1896 and gained postgraduate experience in a number of institutions, among them, the Institut Pasteur in Paris. Back home, in 1907, he was awarded the title of Professor. For many years he worked in what is today the Kovalenko Institute. At the beginning of the 20th century, he earned international fame for his work in horse-breeding using his perfected technique of artificial insemination.

In the mid-1920s, he was again in the Pasteur Institute and he received permission to conduct human/ape hybridization experiments in the French Guinea. The Soviet science administration supported his project financially. First, he inseminated female chimpanzees with human sperms. The experiments were not successful. When he left Africa, he took with him dozens of animals for a new research venue in Sukhumi in Soviet Central Asia. The plan was to inseminate five women volunteers with orangutan sperm, but the animal designated for the experiment died before the experiment could be carried out. Soon, Ivanov's governmental sponsors fell under political prosecution and Ivanov himself became a target of political criticism. He was arrested in 1930 and exiled for five years to Alma Ata (today, Almaty, Kazakhstan). He was allowed to continue his professorial activities in his exile and he died in 1932.

Ivanov's impact can be judged by the fact that the Nobel laureate Ivan P. Pavlov wrote his obituary. Ivanov's story served as inspiration for a novel by Aleksandr Starchakov. On the basis of the novel, Dmitry Shostakovich who had visited Ivanov in Sukhumi, composed an opera, which, however, remained unfinished. Nonetheless, after its manuscript had been discovered in 2004, it was premiered in 2011 in Los Angeles.

Sergei N. Vyshelessky (1874–1958) was a graduate of the Warsaw Veterinarian Institute. He participated in the elimination of infectious animal diseases in many regions of Imperial Russia. He was a visiting scientist in Leipzig in the early 1910s and again in Germany and in Denmark in the mid-1920s. He directed a number of veterinary institutions in the Soviet Union.

Konstantin I. Skryabin (1878–1972) graduated in 1905 from the Yurevsky Veterinarian Institute (today, Tartu, Estonia). He worked at a number of venues before moving to Moscow in 1920. For the next 11 years, he was one of the leading researchers at what is today the Kovalenko Institute. In his subsequent career Skryabin further specialized in microbiology and parasitological research and occupied responsible positions in research institutes and national organizations. He founded a dynasty in microbiology and his son and grandson followed in his footsteps.

Yakov R. Kovalenko (1906–1980) was a microbiologist and veterinarian, a graduate of the Moscow Veterinarian Institute. He was the director of the Institute between 1955 and 1977. The main area of his interest was the infectious diseases of agricultural animals.

Statue of Kliment Timiryazev (S. D. Merkurov, 1923). In WWII, shock waves from a nearby German bomb brought down the statue, but it was put back together within hours.

6

Timiryazev Academy

Considering the history of Soviet agriculture and the excellence of its best scientists, seldom has so much excellence been manifested in so little results. The Timiryazev Academy — officially, the Russian State Agrarian University–Moscow Timiryazev Agricultural Academy — celebrated its 150th anniversary in 2015. Initially its name was Petrovsky Academy after Peter the Great. It was the Moscow Agricultural Institute between 1889 and 1917 though many still liked referring to it as Petrovsky Academy.

We visited the Academy in fall 2016 and its beautiful campus is full of memorials — statues, busts, memorial plaques, and buildings of history. The Timiryazev campus is enormous and our presentation follows a geographically rational route to cover as much ground as possible with reasonable time and effort. The nearest metro station to the campus is the Petrovsko-Razumovskaya (rather than the preceding Timiryazevskaya station, if traveling from the city center).

The main building of the Timiryazev Academy is in the place of the former Petrovsko-Razumovsky mansion. The Russian State bought the estate from its owners to establish here a new institution of higher education for agriculture and for other agricultural institutions. The main building will figure at the end of our tour.

Listvennichnaya Avenue.

Listvennichnaya Avenue leads from the metro station Petrovsko-Razumovskaya to the main building of the Timiryazev campus. Its length is 1300 meters (0.8 miles) and there are 1200 larch-trees (*Larix*) along this road. This tree embodies agricultural culture in Russia. Farmers prepared the most diverse agricultural instruments from this tree. The larch-trees along the Avenue were planted by the director of the Academy, Nikolai I. Zheleznov, and by the chief gardener of the institution, Rikhard I. Shreder in 1863–1866. They used seeds of 1861, the year of the legislation about the emancipation of serfs. This symbolism stressed the importance of this legislation. Throughout the history of the Academy, all important events have gone through Listvennichnaya Avenue.

A tractor and a truck as memorials erected in 2005 in front of 7 Listvennichnaya Avenue.

The main building of what used to be the Goryachkin Agro-engineering University founded in 1930 is at 7 Listvennichnaya Avenue. From 2014, it is part of the Timiryazev Academy. The tractor and the truck were important instruments at the time of WWII. They symbolized the work the associates and students carried out in the home front and in the front itself, respectively. The year 2005 when the two memorials were erected signified the 60th anniversary of the ending of the war. Both the tractor and the truck are claimed to be in working condition and could be put back into operation if the need arose.

Vasily N. Boltinsky (1904–1977) has a memorial tablet on the façade of 7 Listvennichnaya Avenue (not shown). He was an agricultural engineer, an innovator of tractors, and participated in the development of about 25 different models. He involved his students in his design work.

Bust of Vasily P. Goryachkin (N. I. Rudko, 1972) in front of 7 Listvennichnaya Avenue.

Vasily P. Goryachkin (1868–1935) was a disciple of Nikolai Zhukovsky (Chapters 2 and 4) and became a pioneer of agricultural machines. He was head of the department of agricultural machines at the Academy between 1896 and 1930; this is today the department of technology and machines of plant cultivation.

37 Pryanishnikov Street was the cradle of plant genetics in Russia from 1903. Two memorial plaques on its facade honor plant breeders of two generations, Sergei I. Zhegalov (*left*) and Lev A. Trisvyatsky (*right*).

Petr I. Lisitsyn (1877–1948) has a memorial tablet (not shown) on the façade of 37 Pryanishnikov Street. He was a plant geneticist, graduated in

1902 from Moscow University where Timiryazev was one of his professors. Lisitsyn continued his education at the Agricultural Academy. Before he completed his studies, he was exiled for revolutionary activities. From 1908 he worked at an experimental agricultural station and from 1929 he was professor and head of the department of botany at the Timiryazev Academy.

Renowned plant breeders and plant geneticists worked at 37 Pryanishnikov Street, among them Nikolai Vavilov (Chapter 1). The building now houses the Nikolai Vavilov Museum. Here is also the department of plant genetics and biotechnology. Trofim Lysenko (Chapter 1) used to have a cabinet in this building.

Left: The Mikhelson building with an observation tower at its top at 12 Pryanishnikov Street. *Right*: Inside, there is a bust of Vladimir Mikhelson (Robert Bakh[1]).

The Mikhelson building houses the department of meteorology and climatology and the Mikhelson Meteorological Observatory, one of the two oldest meteorological observatories in Russia. The tower at the top is for the meteorological observations, including the measurement of wind strength and the monitoring of solar radiation. Observations have been conducted since 1879.

---

[1] Sculptor Robert Bakh was a descendant of the composer Johann Sebastian Bach.

Vladimir A. Mikhelson (1860–1927) was a physicist and geophysicist, meteorologist, and one of the founders of the science of solar activities in Russia. He graduated from Moscow University in 1883 where he continued his studies and also abroad. After having acquired his doctoral degree he joined the Academy. Mikhelson was the first who applied statistical physics in the investigation of black body radiation. He studied the Doppler Effect and other phenomena in connection with the propagation of waves. He demonstrated the importance of meteorology in agriculture.

Two memorial plaques on the Mikhelson building. *Left*: Vladimir Mikhelson. *Right*: Yury I. Chirkov who worked here between 1970 and 1988 and applied meteorology in agriculture.

3 Listvennichnaya houses the dean's office of the Faculty of Agronomy and it is also a study hall. There is a bust of Ivan A. Stebut (L. N. Matyushin, 2005) in the middle of its yard.

Ivan A. Stebut (1833–1923) complemented his studies in Russia and in Western Europe, mostly in Jena. He joined the Academy in 1868. He founded an agricultural museum, an experimental field, and a laboratory for agronomy. He organized a consultative commission, started a journal, and initiated an agricultural exhibition in Moscow. He contributed to the reorganization of Russian agriculture following the emancipation of serfs and advocated the importance of giving women the possibility of education in agricultural sciences. He and Dmitry Pryanishnikov (see later in this chapter) initiated agricultural courses for women.

The façade of the building at 3 Listvennichnaya displays a number of memorial plaques. We describe briefly those whom the plaques honor along with the images of six of the plaques (plaques with text only are not shown).

From left to right: Petr I. Lisitsyn (see above), Nikolai A. Maisuryan, and Petr P. Vavilov.

Aleksei G. Doyarenko (1874–1958) specialized in physics for agriculture and worked here until his arrest in 1930 on the pretext of anti-Soviet activities. He may have been involved in supporting a political party different from the communist party. Allegedly, he was to become minister of agriculture in a new government if they had succeeded in bringing down the one-party Soviet system. He was sentenced to five years in prison and when it was served, he was exiled for four more years. Afterwards he could resume his professional activities though not in Moscow.

Nikolai A. Maisuryan (1896–1967) was a biologist and head of the department of plant production. He worked at the Timiryazev Academy from 1928 throughout his entire career.

Petr N. Konstantinov (1877–1959), plant geneticist, was professor of plant production from 1936.

Petr P. Vavilov (1918–1984; no relation to Nikolai Vavilov) was a plant geneticist with interest in animal fodder products, especially under the severe conditions of the far northern climate. He advocated the dissemination of the pernicious plant pigweed (*Heracleum spondylium*) or brank-ursine (*Heracleum spondylium sibiricum*). He believed that these plants would help in regenerating the soil. Instead, they turned out to be harmful for the cultivation of useful plants. The forced utilization of these plants did not stop at the borders of the Soviet Union, but spread into Eastern European countries as well.

Nikolai G. Andreev (1873–1932) specialized in meadowland economy and cultivation and was head of the department of meadow science.

From left to right: Sergei A. Vorobiev (1904–1992) professor of agronomy; Anatoly I. Puponin (1940–2000) soil scientist, rector of the Academy and deputy minister for agriculture; and Ivan S. Shatilov (1917–2007) biologist and agronomist, deputy minister for agriculture.

There are many Lenin statues in Moscow, yet it was unexpected to find one at the Timiryazev campus. This statue used to stand in the Leningradsky railway station and was donated to the Timiryazev Academy in 1981. The head of the statue is turned in the direction of the Faculty of Economics which was supposedly organized with Lenin's active support in 1919. The grateful Academy named Lenin an honorary student. Thus the explanation for the statue is that it remembers the honorary student Lenin rather than the politician.

On the façade of the Faculty of Economics and Finances, there are two memorial plaques honoring two repressed scientists.

The economist and sociologist Aleksandr Chayanov.

Aleksandr V. Chayanov (1888–1938) graduated from the Academy in 1911; his mother was among the first women graduates of this institution. Nikolai Vavilov was one of Chayanov's fellow students. Chayanov completed his education in Western Europe. He advocated the merits of cooperatives and participated in the February 1917 revolution. He was appointed professor at the Academy in 1918. Already in 1926, he was accused of anti-Marxist interpretation of the agricultural policy and in 1930 he was arrested for alleged anti-Soviet activities. He was sentenced to 5 years of prison in 1932. His sentence was eventually modified to exile and he could work in his profession in Central Asia. In 1935, his exile was extended by three years. In 1938, he was arrested again, and executed. In 1987, the Supreme Court of the Soviet Union determined that everything was illegal what happened to Chayanov, including the methods — obviously, torture — used to extract from him the admission of guilt.

The economist Nikolai Kondratiev.

Nikolai D. Kondratiev (1892–1938) created the theory of economic cycles and gave theoretical foundation for the New Economic Policy (NEP) in the early 1920s. The communist methods of governing the economy had proved disastrous and the NEP meant their temporary relaxation. Kondratiev was active in politics and taught in several institutions, including the Academy. He was already arrested in 1922 for anti-Soviet activities and was considered for exile from the Soviet Union — such a punishment would have saved his life but it did not happen. His and Chayanov's teachings, in which the usable features of both planned and market economies figured, were criticized at the highest level. Stalin considered their views to represent an anti-Soviet movement.

The NEP lasted until about 1928. Kondratiev was arrested again in 1930 and he was sentenced to 8 years in prison in 1932. In 1938, he was sentenced to death and executed on the same day. He was first rehabilitated in 1963, but the decision about it remained classified and so were his views. His complete rehabilitation came about in 1987.

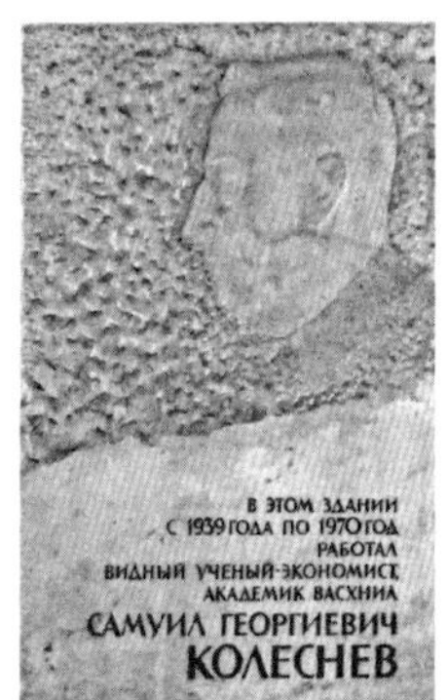

From left to right: Samuil G. Kolesnev (1896–1970); Grigory M. Loza (1907–1981); Sergei S. Sergeev (1910–1999); and Mikhail I. Sinyukov (1924–1996). They were agricultural economists; held professorships and high administrative positions.

There are four memorial plaques inside the building of the Faculty of Economics and Finances. There are memorial tablets for three more economists on the façade of the building at 1 Verkhnaya Avenue: Vladimir A. Dobrynin, Gennady I. Budylkin, and Evgeny B. Khlebutyn.

Nikolai Vavilov was a student at the Timiryazev Academy and started his career at his Alma Mater.

Statue of the student Nikolai Vavilov (L. N. Matyushin, 2015) on Listvennichnaya Avenue.

Nikolai Vavilov was having a spectacular career until it started eroding and ended in a most tragic way (Chapter 1).

*****

In WWII, the associates and students of the Academy made heroic sacrifices and several memorials pay tribute to them on the campus.

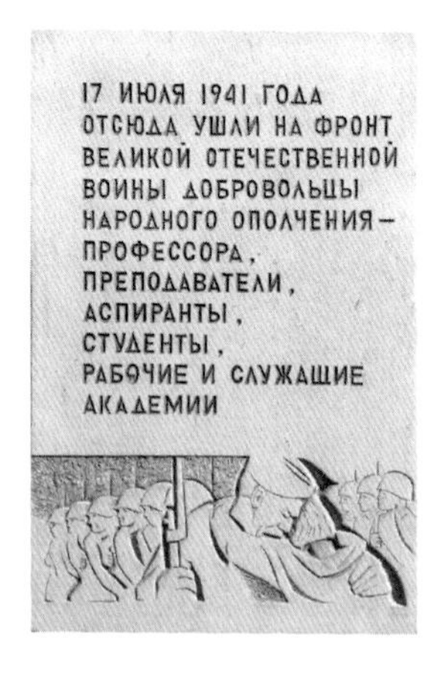

Left: On July 17, 1941, volunteers — professors, instructors, doctoral students, students, workers, and staff members of the Academy — departed to the front. *Right*: In remembrance of the fighting battalion formed from professors, instructors, doctoral students, students, workers, and staff members of the Academy in WWII, 1941–1945.

*****

Busts of Nikolai Zheleznov (L. N. Matyushin, 2006) and Rikhard Shreder (L. N. Matyushin, 2012) at the beginning of Listvennichnaya Avenue, on its opposite sides.

Nikolai I. Zheleznov (1816–1877) was a botanist and agronomist and a professor at Moscow University. His main interest was in the physiology of plants. He founded the Petrovsky Academy and was its first director. He was one of Timiryazev's mentors. His bust stands close to the science library, which he founded using his own money.

Rikhard I. Shreder (1822–1903), the Danish-born "father of Russian gardening," was court gardener to Aleksandr II and came to the Academy at Zheleznov's request. Shreder reconstructed the neglected parks and tree nurseries; he was an exceptional landscape architect.

Bird's eye view of the central part of the campus (courtesy of Stanislav Velichko). The main building of the Academy is in the center of this image (at the intersection of the two diagonals of the image rectangle). Timiryazevskaya Street runs in its front, parallel to the building. Listvennichnaya Avenue runs perpendicular to the building. All four corner towers of the building at 48 Timiryazevskaya Street are seen towards the lower left segment of the image.

The building at 48 Timiryazevskaya Street used to belong to the agricultural farm of the Razumovsky family. Now it houses a veterinarian clinic, the dean's office of the Faculty of Zoo-engineering and Biology, the department of milk and meat, and the only museum of horned cattle in Russia. This building is the oldest on the campus where the activities of the Academy began in 1865. In 1812, Napoleon Bonaparte met with his generals in this building.

At 48 Timiryazevskaya Street. *Left*: Memorial plaque of Vasily S. Shipilov (1924–1991), a specialist in veterinarian obstetrics. *Right*: A hall in the Museum of Animal Husbandry (courtesy of Stanislav Velichko) with a bust of Mikhail F. Ivanov (1871–1935), a specialist in animal husbandry.

The building at 44 Timiryazevskaya Street houses rich contents, among them the Kulagin Zoological Museum, the Gindtse Museum of Anatomy, and the Museum of Horse-breeding.

Left: The front of 44 Timiryazevskaya Street. *Right*: The entrance lobby with a war trophy (see text).

Two memorial plaques in the entrance lobby. *Left*: Nikolai Kulagin. *Right*: Boris A. Kuznetsov (1906–1979), head of the department of zoology between 1956 and 1979.

Nikolai M. Kulagin (1860–1940) was a zoologist and head of the department of zoology between 1894 and 1940. In the 1930s, he was a member of the committee charged with overseeing nature conservation areas — today, we would call them national parks.

Boris K. Gindtse (1881–1951) moved to the Academy from the First Moscow Medical Institute. He founded a museum to facilitate the instruction of students. His primary interest was in the arterial systems of blood circulation in the brain and the collection of the museum is especially rich in exhibits related to this topic.

The horse sculpture standing in the entrance lobby was a war trophy brought back from Eastern Prussia at the end of WWII. It is said that the former cavalry officer, later Marshal, Semen M. Budenny (1883–1973) took fancy of the sculpture and this is how it got onto the list of items for war reparation.

Four scenes in the Museum of Horse-breeding (courtesy of Stanislav Velichko).

Marshal Budenny helped create the Museum of Horse-breeding whose rich collection is based on the legacy of Yakov I. Butovich (1881–1937). Butovich trained as a cavalry officer and distinguished himself in the Russo-Japan war. He then continued his studies of animal husbandry at European universities and specialized in horse-breeding. He developed a renowned farm and his horses won prestigious prizes. After the Bolshevik revolution, the farm was nationalized, but Butovich was allowed to remain in its charge. He did much to save horses nationwide during the ensuing civil war and became a consultant to the legendary cavalry commander Semen Budenny. In the 1920s, Butovich was persecuted for being a burgeouse specialist, but only in 1933 was he arrested. When he was freed, he was barred from living in some of the major cities of the Soviet Union. At this time he cataloged his former extraordinary collection of artifacts about horses. In 1937, he was arrested again and at this time he was sentensed to death as an "enemy of the peoople," and was executed. In January 1989, he was rehabilitated from the 1937 sentence and in March 1989, also from the 1933 sentence.

Nikolai Khudyakov's bust (Robert Bakh, 1928) and his memorial tablet (2016) in the lobby (courtesy of Stanislav Velichko.) at 50 Timiryazev Street.

Nikolai N. Khudyakov (1866–1927) was a microbiologist and plant physiologist who originated from nobility. His physician father was a general in the army medical service. Khudyakov studied at the Petrovsky Academy and at the universities of Berlin and Leipzig. He held a professorial appointment at Moscow University when the controversy with the minister of education arose in 1911. He left in protest together with many other professors. One of Khudyakov's few publications was about agricultural microbiology, which then served generations of microbiologists. He and his colleagues discovered the phenomenon of adsorption of bacteria on particulates of the soil.

Memorial plaque of Elly A. Bogdanov (1872–1931; G. A. Ognev, 1972), a specialist in animal husbandry, on the façade of 52 Timiryazev Street, housing the department of animal husbandry.

Bust of Kliment Timiryazev (Mariya M. Strakhovskaya, 1924) in Petrovsky Square, opposite the main building of the Academy. As people are leaving the main building and look in the direction of the bust, their eyes meet with Timiryazev's eyes.

Kliment A. Timiryazev (1843–1920) was a naturalist with his principal interest in plant physiology and photosynthesis. He popularized science and was a science historian. He was among the first supporters of Charles Darwin's teachings.

Timiryazev was born into a family of nobility, which was British on his maternal side. On his travels to Great Britain, he met with Darwin. Timiryazev was a student of St. Petersburg University and conducted agrochemical experiments at Mendeleev's instructions. Some of his other professors were also distinguished representatives of French and German science, such as Marcellin Berthelot, Robert W. Bunsen, Hermann Helmholtz, and Gustav Kirchhoff. Timiryazev started publishing in 1868 and his professional activities started at the Petrovsky Academy in the early 1870s. He was appointed professor at Moscow University in 1877; left the University in 1911 in protest against the violation of university autonomy, and resumed his appointment in 1917. He was a co-founder of the Higher Courses for Women in Moscow.

Timiryazev welcomed the Bolshevik revolution and the Soviet regime expressed its appreciation to Timiryazev by naming streets and institutions after him. His critical, though not at all opposing, approach to Gregor Mendel's teachings of inheritance was later blatantly misused by Trofim Lysenko in his unscientific and anti-science activities.

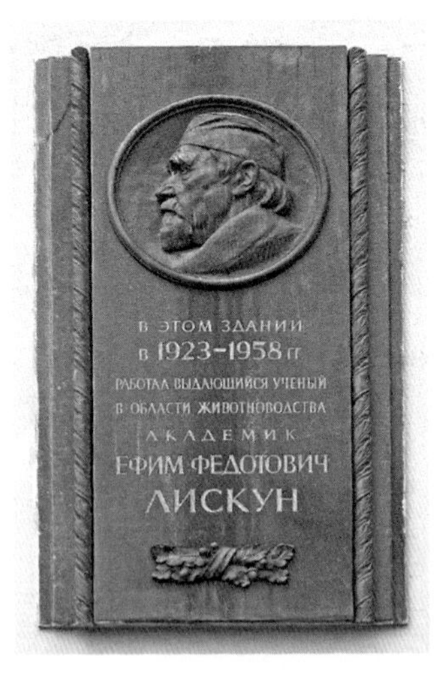

Memorial plaques on the façade of 54 Timiryazev Street. *Left*: Efim F. Liskun (1873–1958), professor of animal husbandry. *Right*: Ivan S. Popov (1888–1964), professor of animal fodder. A modest tablet honors Pavel N. Kuleshov (1854–1936), professor of zoo-technology.

2 Timiryazev Passage (designed by N. N. Chernetsov, 1912–1914) houses the Faculty of Agro-chemistry, Soil Science, and Ecology.

The building at 2 Timiryazev Passage was a gift from Humboldt University of Berlin to mark the centennial of the 1813 "Battle of the Nations" at Leipzig. Originally, it was to accommodate chemical laboratories and professorial apartments. The design of the building addressed two additional goals. One was that the rotunda of the building was shaped to help disseminate the sound of the alarm bell all over the campus — the fire station was situated opposite 2 Timiryazev Passage. The other was to have an auditorium with excellent acoustics for concerts and readings of poetry. There are four memorial plaques on the façade of the building.

From left to right: Evgeny N. Gapon, Nikolai Ya. Demiyanov, Ivan A. Kablukov, and Gavriil G. Gustavson.

Evgeny N. Gapon (1904–1950), professor of physical chemistry and soil scientist, cooperated with the nuclear physicist Dmitry D. Ivanenko of Moscow University. Nikolai Ya. Demiyanov (1861–1938), professor of organic chemistry, synthesized new compounds and investigated intramolecular changes. Ivan A. Kablukov (1857–1942), professor of inorganic and analytical chemistry, researched artificial fertilizers and solutions, and cooperated with the Swedish Svante Arrhenius and the German Wilhelm Ostwald. Gavriil G. Gustavson (1842–1908), professor of organic and agronomic chemistry, researched catalysis of organic synthesis and taught also at the Higher Courses for Women.

The memorial of Mitrofan K. Tursky (P. V. Dzyubanov, 1912) stands adjacent to 60 Timiryazev Street. The relief depicts Leo Tolstoy planting a tree with a child looking on.

Mitrofan K. Tursky (1840–1899) placed forestry onto scientific foundations. From his time the tradition has developed that every freshman and every senior upon graduation plants a tree or a bush on the campus. There is a special day for this, which has become a popular holiday for everyone at the Academy. Tursky and the writer Leo Tolstoy were great friends. Tolstoy often visited the Academy and the famous author was happy to interact with the students. For Tursky's memorial a collection was made among forestry institutions in Russia. More money came together than needed for the memorial and the unused sum was donated to plant forests in Moscow suburbs.

Left: Vargas de Bedemar Museum of Forestry at 13 Pryanishnikov Street. The Danish-born Alfons R. Vargas de Bedemar (1816–1902) was a specialist in the economics of forestry and tree taxonomy. *Right*: Book-binders produced from tree rinds in the Museum of Forestry (courtesy of Stanislav Velichko).

From left to right: Bust of Aleksei N. Kostyakov (1887–1957; N. B. Nikogosyan, 1976) in front of 19 Pryanishnikov Street. The building houses the Kostyakov Institute of Soil Melioration and Hydro-construction. The memorial plaque honors Sergei F. Averyanov.

Sergei F. Averyanov (1912–1972) hydrologist and soil scientist participated in WWII; from enlisted man he made it to engineer-captain. In 1943 he suffered a serious wound and lost a leg. He returned to Moscow in 1944 and continued his teaching and research at the Timiryazev Academy. During the last decade of his life he was head of the department of agricultural soil science. His achievements in the research of soil-amelioration earned him international recognition.

The building at 6 Pryanishnikov Street houses the Faculty of Gardening and Landscape Architecture; there are six memorial plaques on its wall.

Two large memorial plaques on the façade of 6 Pryanishnikov Street. *Left*: The plant geneticist and specialist of vegetable culture German I. Tarakanov (1923–2006; G. V. Frangulyan, 2015). *Right*: Vitaly I. Edelshtein (1881–1965), specialist of olericulture — the science of vegetable growing.

Memorial plaques on the façade of 6 Pryanishnikov Street, *from left to right*: pomologist Petr G. Shitt (1875–1950); plant geneticist Nikolai N. Timofeev; vine cultivator Aleksandr M. Negrul; and pomologist Benedikt A. Kolesnikov (1895–1978).

Statue of Dmitry N. Pryanishnikov (G. A. Shults and O. V. Kvinikhidze, 1973) on the small grassy square adjacent to 6 Pryanishnikov Street.

Dmitry N. Pryanishnikov (1865–1948) graduated in 1887 from Moscow University where Markovnikov (Chapter 2), Timiryazev, and Stoletov (Chapter 2) were among his professors. Pryanishnikov joined the Petrovsky Academy and soon embarked on a two-year study in Western Europe. He investigated the metabolism of proteins and other nitrogen-containing substances in plants. Later, his primary research area was the application of fertilizers and plant physiology. From 1895 for the rest of his life he was head of the department of agro-chemistry at the Academy. He was the founder of agro-chemistry in Russia. He also taught at Moscow University.

Detail of the memorial plaque of Vsevolod Klechkovsky (G. A. Shakarov, 1978) on the façade of a wing of 6 Pryanishnikov Street.

Vsevolod M. Klechkovsky (1900–1972) was an agro-chemist with deep interest in the fundamentals of structural chemistry. He pioneered in Russia the application of radioactive isotopes in the investigation of metabolism in plants. There is another memorial tablet next to the one honoring Klechkovsky. According to this tablet, in 1947, a biophysical laboratory was organized at the Timiryazev Academy in the framework of the Soviet nuclear program. Here, as a first in the Soviet Union, they conducted radiological research for agriculture.

The archival photograph shows the greenhouse at the 1896 Exhibition in Nizhny Novgorod. After the Exhibition, Timiryazev had the greenhouse transported to Moscow as a gift to his teacher, Dmitry Pryanishnikov. Courtesy of Stanislav Velichko. Part of the metallic roof of Timiryazev's greenhouse is preserved in front of a wing of 6 Pryanishnikov Street.

A wing of Pryanishnikov Street 6 dates from the 18th century and from the beginning of the Academy there was a museum of agriculture and the department of botany at this venue. Timiryazev, Vasily Vilyams (see, below), and Pryanishnikov used to work here and Nikolai Vavilov and Aleksei Doyarenko (see, above) studied here. Currently sections of the Faculty of Agro-chemistry and the Faculty of Soil Science operate in this building.

Statue of Vasily R. Vilyams (S. O. Makhtin, 1947) at 49 Timiryazevskaya Street.

Vasily R. Vilyams ([Williams], 1863–1939) was one of the founders of soil science in Russia. His father immigrated to Russia from the United States. Vasily Vilyams studied at the Petrovsky Academy and upon graduation he stayed on and at once was sent for a study trip to France and Germany and later to the United States and Canada. He started teaching in 1891. He initiated a nursery garden in 1904 in which he collected species of cereals and leguminous plants. In another initiative he established an experimental station for studying animal fodder. From this, the current Vilyams Fodder Research Institute has developed.

Vilyams followed up Vasily Dokuchaev's teachings in turning large plains often suffering from the worst draught in Russia into fertile areas. Dokuchaev proposed to regulate rivers, establish water-basins, and forestation for improving the rainwater economy. Vilyams declared that Soviet science can turn any soil into a high-producing one and he worked out a grass-arable rotation system. He observed that fallow land can be restored by planting certain grasses. Such an approach may prove workable, but only if a lengthy period is allowed. Vilyams wanted to shorten this period by alternately planting grasses and perennial papilionaceous plants that would loosen the soil and enrich it in nitrogen. He complemented this with forest belts that would shelter the soil from the drying hot winds, help balancing the climate in the micro-regions, and restrain the rainwater from quick departure.

At Vilyams's time, there was a scarcity of chemical fertilizers and agricultural machinery, and this approach could have helped on the short run though might have hindered progress of agriculture in the long term. What was worse that this approach was recommended, even forced, not only for restoring fallow land but everywhere, including the East-European countries that had become parts of the Soviet sphere of influence following WWII (by then, Vilyams had been dead).

There was a dark side of Vilyams's career. He conducted sharp debates with other scientists and in the mid-1930s such disagreements could lead to tragic consequences. Thus, when he had a disagreement concerning certain techniques in farming with academician Nikolai M. Tulaikov (1875–1938), Tulaikov was arrested and perished. Vladimir Vernadsky wrote in 1943 that Vilyams was leaving behind a bad and distorted school and that he used inaccurate material, which sometimes contradicted reality. According to Vernadsky, Vilyams having joined the [communist] party did not make him an authority, and Vernadsky predicted that Vilyams would be soon forgotten.[2] Even though Vernadsky's assessment of Vilyams's teachings appears to be valid, judging by the memorials honoring him he is not forgotten.

The main building of the Timiryazev Academy and a memorial plaque of Nikolai I. Vavilov (A. M. Balashov, 1989) on its façade. Nikolai Vavilov studied and worked here between 1906 and 1917.

The main building as it stands today is not the original mansion. It was replaced by what the designers of the campus in the 1860s thought would better serve the educational institution. Architect Nikolai L. Benua prepared the design and architect P. S. Kampioni led the construction. In addition to the Nikolai Vavilov memorial, there are two more memorial tablets on the façade: one remembers the former Soviet figurehead president Mikhail I. Kalinin and the other the Russian writer, and former student of the Academy, Vladimir G. Korolenko.

[2] https://ru.wikipedia.org/wiki/Вильямс,_Василий_Робертович; accessed 11/17/16.

In the great entrance hall of the main building there are memorial tablets and busts.

Plaque with the portrait of Aleksandr II (Ashot Sardaryan) refers to the event of December 3, 1865, when he presented the Petrovsky Academy its Constitution.

Bust of Kliment Timiryazev flanked by two marble tablets in the great entrance hall. Courtesy of Stanislav Velichko. *On the left*: Extract from a 2009 Resolution of the heads of governments of the Commonwealth of Independent States (CIS)[3] according to which the Timiryazev Academy is a central institution of all participants of CIS in the preparation, improvement of qualification, and re-training in the field of agrarian education. President Vladimir Putin signed it for the Russian Federation. *On the right*: President Dmitry Medvedev (2008) declares the Timiryazev Academy to be especially valuable for the cultural heritage of the Russian Federation.

[3] Following the break-up of the Soviet Union, 12 of its 15 republics (the Baltic States were the exception as they were joining NATO) formed a loose alliance, the CIS.

Busts of Dmitry Pryanishnikov and Nikolai Vavilov in the great entrance hall.

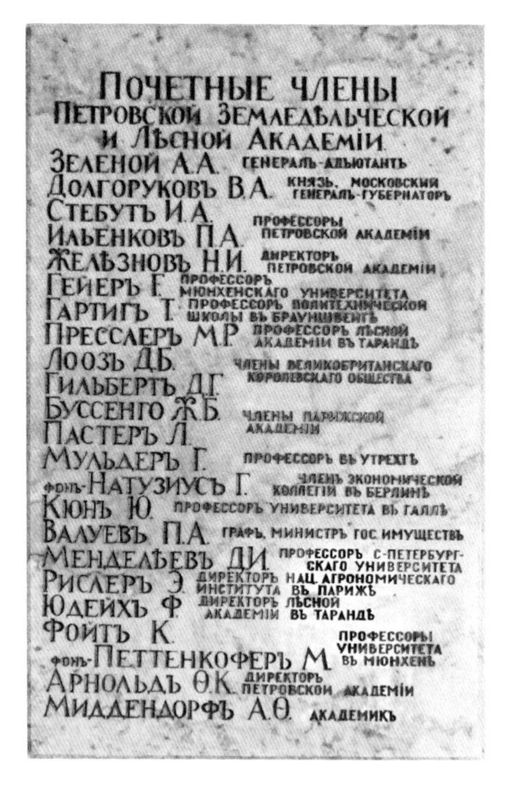

Marble plaque with the list of Honorary Members of the Petrovsky Academy. Two among the impressive list of names: Louis Pasteur and Dmitry Mendeleev.

The grandiose staircase of the main building of Timiryazev Academy.

Upon the completion of this grand tour of the Timiryazev campus, we were duly impressed by the magnitude, variety, and depth of the potential and the accomplishments of this institution. At the same time, looking back to the turbulent history of Soviet politics and science, especially the science connected with agriculture, an uneasy feeling is forming. The question is whether our largely sterile presentation of science memorials on the Timiryazev campus might create a false impression of peace and calm as far as the past is concerned. There were two memorial plaques on the façade of the Faculty of Economics and Finances that honored repressed scientists. There was also the comment we made about the tragic outcome of the scientific debate between Vilyams and Tulaikov. There is then the statue and the memorial plaque of Nikolai Vavilov, but they specifically refer to the years of his studies and the beginning of his career without a hint of his tragic fate. Vavilov left the Academy in 1917 and between that time and the collapse of the Soviet Union so much happened that it is conspicuous how silent the memorials are about it.

Winter view of the back of the main building of the Timiryazev Academy. Courtesy of Stanislav Velichko.

In Chapter 1, we have already touched upon the tragic consequences of Trofim Lysenko's reign in Soviet biology and agriculture and its direct impact on the life and activities of the Timiryazev Academy. The infamous August 1948 session of the Lenin Academy of Agricultural Sciences dealt the Timiryazev Academy a near-fatal blow. Entire directions of research stopped, such as genetics of plants, animals, and fodder; morphology; agrarian economics; and others. Many of the professors of Timiryazev Academy were fired, among them the rector, V. S. Nemchinov. Nemchinov was an economist rather than a biologist; he bravely made a stand for the science of genetics. Nationwide, thousands of biologists lost their jobs.

In tragic irony, Trofim Lysenko was the head of the department of genetics, and, simultaneously, he was the director of the Institute of Genetics of the Academy of Sciences. His lectures were standing room only. Two factors fed their popularity. Students from the villages with a weak high school education liked Lysenko's lectures for their simple language and straightforward contents that was easy to absorb. Lysenko's presentations were unsophisticated to the extent that many other students also found them entertaining for that very reason. The conflict between the folksy academician and the other biologists might have been considered humorous had it not had so tragic and devastating consequences for human lives and for the entire Soviet agriculture.

From the 1930s through the 1950s and beyond, the Timiryazev Academy was not much the venue of front-line science and the institution of higher education of future scientists. Rather, it was turned into a school for preparing administrators for the Soviet cooperatives and state agricultural plants. It also bore the arrogance of the high government bureaucrats with often no higher than middle-level schooling who considered themselves better suited to make decisions about agriculture than the scientists. Although in words the Stalinist regime welcomed debates, in reality the regime ruthlessly repressed those whose opinions differed from what it considered correct. Often views and practical approaches won over scientific conclusions when they promised quick results even though they never became reality. This is why Lysenko could cling to his power not only under Stalin, but also under Khrushchev.

In 1974, years after he had been removed from his influential positions, Lysenko still clung to his unscientific views. He wrote, "In our work, we have not applied and have no intention to apply any ideas or techniques of molecular genetics, whatsoever. I would like to advise all biologists, geneticists, and the students of the Soviet Union not to accept these ideas and techniques, because they hinder the understanding of the essence of the living, i.e., the progress of theoretical biology ..."[4] By then, however, he could only "advise," and no longer use force to make people accept his views.

It was a direct recognition of the failure of Soviet agriculture when under Leonid Brezhnev the Soviet Union was exporting natural gas and oil and importing bread. Mind-boggling as it is but even in the most critical times of food shortage the law persecuted people who cultivated vegetable and fruit gardens; small farmers; clever peasants who loved to work on however small piece of land they could. Part of the paranoia was the fear of letting western ideas into the country even if this meant food shortages if not outright starvation. The agricultural scientists were led to understand that their discoveries and innovations not only were not needed; they might become harmful to their personal well-being. By the 1980s, deviation from ideological purity — whatever the party line was — no longer threatened their lives, but could still be highly disadvantageous. The permanent state of collapse of the Soviet agriculture preceded and contributed to the collapse of the Soviet Union.

[4] N. V. Ovchinnikov, *Akademik Trofim Denisovich Lysenko* (in Russian, Moscow: Luch, 2010), p. 188.

The gravestone of the legendary explorer Otto Schmidt at the Novodeviche Cemetery with the Novodeviche Monastery in the background and the grave of an army general in between.

7

Novodeviche Cemetery

The Novodeviche Cemetery, adjacent to the Novodeviche Monastery, is the most famous among the Moscow cemeteries. It is a cultural-historical memorial under UNESCO protection.[1] It is at 2 Luzhensky Avenue in the southwestern section of the city, relatively small, but is in the heart of the metropolis. From the minute it opens in the morning, busloads of visitors swarm through its gates without stopping until its closing hour in the evening.

The Novodeviche Monastery was founded in 1524 with a small cemetery as part of the monastery, which was filled by the end of the 19th century. An area of 2.7 hectares was added in 1904 as a new section, walled off from the southern direction of the monastery. Burials were conducted according to the Eastern Orthodox religion until 1918. The name of the cemetery became Novodeviche Cemetery in 1919 and from then on, only exceptional people could be buried here, and this exclusivity was formalized in 1927. Some important graves were transferred from other cemeteries, not only politicians but also artists and writers. The transfers offered an opportunity for altering some tombstones in order to adjust them to the Soviet conditions. For example, the cross was removed from the grave of the Russian writer Nikolai Gogol.

Additional areas kept enhancing the territory of the cemetery and in 1949 a wall erected between the existing and added areas accommodated a new set of columbaria. By the beginning of the 21st century, the total area amounted to 7.5 hectares and some 26 thousand people have been buried in it. Once again, the cemetery is full and while existing graves can accommodate additional family members, in principle, no new graves can be opened. When Boris N. Yeltsin (1931–2007), the first president of the new Russia died, a spot had to be carved out of the main walkway of the overcrowded Cemetery for his grave.

The Novodeviche started its operation in the time of the czars, but it became what it is in the Soviet era. Statesmen, artists, writers, military leaders, and scientists and technologists of national and often even international importance have their graves here. If there was ever an expression of the class system in Soviet society (in principle, it was supposed to be classless), this cemetery was an example of it. Descendants could also be buried here, so it resembled hereditary nobility.

---

[1] UNESCO: United Nations Educational, Scientific and Cultural Organization.

The most diverse styles of art intermingle here with the Soviet-favored "socialist realism" dominating. Some of the nation's best artists, sculptors and architects created tombstones for the Novodeviche and then the same sculptors and architects may have won the right to be buried there too.

Nikita Khrushchev's grave by Ernst Neizvestny. The smaller column on the left is for Khrushchev's son, Leonid (1917–1943) killed as a fighter pilot in WWII.

Main walkway with Boris Yeltsin's grave shaped as a Russian tricolor.

The most famous grave at the Novodeviche is Nikita S. Khrushchev's (1894–1971). The sculptor of the memorial, Ernst Neizvestny (1925–2016), had a memorable polemic with Khrushchev in 1962 at an exhibition of contemporary Soviet art. It was unprecedented that an artist — or anybody, for that matter — would enter such a public polemic with a Soviet supreme leader. Neizvestny suffered the consequences; afterwards, he received no commissions and the authorities persecuted him. In a curious twist, when Khrushchev died, his family insisted that Neizvestny create Khrushchev's tombstone. A unique memorial resulted. The Soviet leadership was apprehensive of cultivating Khrushchev's memory and following Khrushchev's funeral, the Cemetery was closed to visitors for years under the pretext of renovation. Then, after the Cemetery reopened in the 1980s, for years, photography was not allowed. Over the years we have visited this cemetery numerous times and have collected images of graves that would fall into the categories covered in this book.

It was a great privilege to be buried in the Novodeviche Cemetery and many privileged scientists in the Soviet system were truly excellent. Many other truly excellent people were left out of the privileged circles. There were scientists of great promise or of already proven merit who perished under Stalin or were just not let into the privileged circles under his successors. Thus, the selected ones whose graves are at the Novodeviche cannot be considered necessarily the top of the top in every instance. On the other hand, there are superb names among them.

We list here the scientists and technologists whose graves we have found at the Novodeviche. Of those, we reproduce the images of some eighty graves; and we add vignettes to some entries, especially if the person did not figure in previous chapters. The importance of the buried and even more the visual interest of the gravestones played a role in our choices for presenting actual images. We divided the entries in the following nine groups and listed the entries in alphabetical order within each group.

Mathematicians and computer technologists
Physicists
Rocket and space scientists and technologists
Airplane designers
Engineers, technologists, and inventors
Chemists and materials scientists
Biologists and agricultural scientists
Medical scientists and physicians
Earth scientists and explorers

For a non-visitor reader this should provide a convenient orientation. Even a visitor might find this approach useful because few visitors would like to visit all the graves of scientists and technologists. This approach helps finding specific graves. A typical science-oriented visitor would probably be more interested in covering graves that belong to certain areas and disciplines. Where we had the data, we quote the identification of the grave by section number (*uchastok*) and row number (*ryad*) as [uchastok/ryad]. However, according to our experience, even these two numbers may not be sufficient to find a grave easily.

Mathematicians and Computer Technologists

Vladimir Arnold [11/6] and Andrei Kolmogorov [10/4].

Vladimir I. Arnold's (1937–2010) gravestone is full of mathematical expressions. The long sentence on his gravestone says in a rough translation: "Being a mathematician, I, all the time, have to rely on sensations, guesses and hypotheses rather than on proofs. I am moving from one fact to another utilizing the special illumination that makes me to consider the common aspects of the phenomena under study. To a bystander, these aspects may not even appear connected to each other."

Arnold started doing math as a teenager, studied at Moscow University, where Andrei Kolmogorov was his mentor, and stayed on after graduation. Arnold worked in several crucial areas of mathematics and solved problems that others had declared unsolvable. He wrote a book about the so-called *catastrophe theory*. It deals with phenomena in which gradual changes suddenly turn into major changes under certain conditions that appear impossible to predict. Arnold divided his last dozen years between Moscow and Paris.

Anatoly Dorodnitsyn [10/8] (Chapter 1).

Andrei N. Kolmogorov (1903–1987) considered theory of probability to be the closest to him, but he worked in many areas of mathematics. He was a pupil of Nikolai Luzin and visited David Hilbert and Richard Courant in Gottingen in 1930. Kolmogorov was appointed professor at Moscow University in 1931 and full member of the Academy of Sciences in 1935 soon after the scientific degree system had been introduced. Kolmogorov worked for the artillery in World War II and returned to his mathematical research after the war. He helped reform high school teaching and developed a strong school of mathematicians. Some of his disciples have continued their careers in the West following the collapse of the Soviet regime.

Nikolai M. Krylov (1879–1955) [3/38] graduated in 1902 from the St. Petersburg Mining Institute, where he later became professor. Eventually, he moved to Kiev and was in charge of the mathematical and physical section of the Ukrainian Academy of Sciences. He published papers in pure mathematics and was one of the initiators of nonlinear mechanics.

Sergei A. Lebedev [3/50] (Chapter 4).

Gury Marchuk [11/7] and Ivan Petrovsky [7/5] (Chapter 2).

Gury I. Marchuk (1925–2013) was both a mathematician and a physicist. His research aimed at applications in a variety of areas, including nuclear reactors, the dynamics of the atmosphere, computation, automation, and regulation. Between 1986 and 1991 he was the President of the Academy of Sciences. The rough translation of the quote on his gravestone is: "Science is the brain center of society. Its impact is not felt immediately, but it is the factor that transforms peoples' lives. Therefore, society should be considerate towards the value science represents."

Lev Pontryagin [10/5] (Chapter 1).

Ivan I. Privalov (1891–1941) [4/23] was a graduate of Moscow University. While still being a student, he attended lectures of David Hilbert, Felix Klein, and other giants of mathematics in Gottingen. From 1922, Privalov held a professorial appointment at his Alma Mater and from 1923 he was section head at the Research Institute of Mathematics and Mechanics as well as professor at the Zhukovsky Air Force Academy.

Sergei Sobolev [10/5] and Ivan Vinogradov [10/1] (Chapter 1).

Sergei L. Sobolev (1908–1989) lost his father early in his childhood. His mother, both a teacher and a physician, was determined to provide the best education for the boy who appeared very gifted. He had a meteoric career in mathematics. In 1929, Sobolev graduated from Leningrad University where internationally renowned mathematicians were among his professors. After graduation, he worked in mathematical physics and in 1934, he joined the Steklov Institute of Mathematics.

From 1945, Sobolev worked for the nuclear project. Parallel to this work, he completed his principal monograph about the fundamentals of mathematical physics. In 1952, Moscow University invited Sobolev for a prestigious professorship at its recently organized department of computational mathematics. Eventually, the Computational Center of the University developed from this department. Sobolev, together with Mikhail Lavrentiev and Sergei

Khristianovich, was among the initiators of the Siberian Branch of the Academy of Sciences. From 1957 until 1983, Sobolev directed the Institute of Mathematics in Novosibirsk.

Sobolev's activities promoting computational technology were the more remarkable, because in the 1950s, cybernetics, that is, computational technology, along with genetics, was considered pseudoscience and ideologically taboo in the Soviet Union. Sobolev already in the mid-1950s fought for reversing this approach and for the recognition of the importance of computational technology. In the mid-1960s, he took a stand in defense of Leonid Kantorovich when Kantorovich was being attacked for applying mathematics in economics, which was declared contrary to Marxism-Leninism.

Vyacheslav V. Stepanov (1889–1950) [4/23] graduated from Moscow University and attended lectures of David Hilbert and Edmund Landau in Gottingen. Stepanov became professor of mathematics at his Alma Mater and stayed there until the end of his life. He was head of theoretical geophysics at the State Astrophysical Institute between 1929 and 1938. In 1939, he was appointed to be director of the Moscow Research Institute of Mathematics and Mechanics.

Physicists

Abram I. Alikhanov [5/38] and Nikolai N. Andreev [7/17].

Abram I. Alikhanov (1904–1970) was one of the initial participants of the Soviet nuclear project. He had a career rich in experimental discoveries. He participated in the first experiments producing artificial radioactivity. In 1945, he co-founded the Institute of Theoretical and Experimental Physics (25 Bolshaya Cheremushkinskaya Street) and remained its director until 1968. Today, this Institute bears his name. In 1945, Alikhanov was assigned to a leading role in the atomic bomb development.

Nikolai N. Andreev (1880–1970) graduated from the University of Basel. Among his most significant venues of employment were Moscow University, the Leningrad Polytechnic Institute, and FIAN. His principal research interest was in acoustics and he founded a scientific school in this area of science. Today, there is an Andreev Acoustics Institute (4 Shvernik Street).

Vladimir K. Arkadiev, (1884–1953) [1/29] at Moscow University had such mentors as Nikolai Umov and Petr Lebedev. Arkadiev visited laboratories in Western Europe before launching his career at home. In WWI, he was engaged in research of defense against chemical weapons. After the war, his principal work was at Moscow University. He was married to the physicist Aleksandra A. Glagoleva-Arkadieva (1884–1945) who is buried in the same grave.

Lev Artsimovich [7/3] (Chapter 1) and Nikolai Basov [11/5] (Chapter 1).

Nikolai Bogolyubov [10/6] (Chapter 2).

Viktor-Andrei Borovik-Romanov (1920–1997) [2/28] fought in WWII and graduated from Moscow University in 1947. He started his career as a physicist in Kapitsa's Institute, 1947–1948, where he returned in 1956. He was appointed director of the Institute after Kapitsa died, from 1984 until 1990.

Pavel Cherenkov [10/5] (Chapter 1). The carving in the stone refers to the Cherenkov Effect.

Evgeny A. Chudakov [1/2] (Chapter 1).

The tombstone of Gregory N. Flerov (1913–1990) [10/16] is a large Eastern Orthodox cross as if compensating for his communist party membership and his Jewish origin on his mother's side. He and his brother were brought up by their mother alone after their father had died. Flerov worked in various plants before he was directed in 1933 to study in Leningrad. He completed his Diploma Work (Master's degree equivalent) in 1938 under Igor Kurchatov's mentorship and stayed with Kurchatov at the Leningrad Institute of Physical Technology.

In 1941, he enlisted, but even under heavy fighting he managed to keep up with the scientific literature. He made a most remarkable deduction that the West must have already started working on the atomic bomb, because those whom he judged capable of taking part in it had disappeared from the scientific literature. In 1942, he wrote to Stalin about the necessity of developing the atomic bomb and about how it could be done. It was at the time when Soviet intelligence was already transmitting information from Great Britain about the preparations for a possible atomic bomb.

Flerov had a brilliant career in the Soviet nuclear program. He directed the work leading to the discovery of new heavy elements and helped found the Joint Institute for Nuclear Research in Dubna. In 1998, scientists in Dubna discovered a new element, No. 114. According to international convention, the discoverers suggest the names for the new elements. This is how it happened that element 114 was named after Flerov, Flerovium, Fl.

Vasily G. Fesenkov, (1889–1972) [Columbarium, Section 135] was an astronomer, one of the founders of astrophysics in the Soviet Union. He was a leading physicist in Kazakhstan.

Vitaly Ginzburg [11/6] (Chapter 1).

Vitaly Ginzburg had his share of difficulties under the Soviet system. In the period between 1946 and 1953 (Stalin's death) both anti-Semitism and the regime's ideological fight against modern science meant life-threatening dangers for him. Also, after his divorce from his first wife, he married an exile, Nina Ermakova, who had been falsely accused of conspiracy to kill Stalin. For many years she was not allowed back to Moscow. In old age, looking back on his exceptionally rich achievements in science, Ginzburg wondered whether easier conditions might have allowed him to perform even better. He came to the conclusion that had he lived under better conditions, he might have been happier and seen more, but the sum of his scientific output might have not been larger than it was under the circumstances he actually had lived his life.

Petr Kapitsa [10/2] (Chapter 1).

The simplicity of Petr Kapitsa's [10/2] (Chapter 1) grave is in full accord with his personality. He was an aristocrat in his demeanor. For example, he did not wear his many medals, except for the two stars of Hero of Socialist Labor. He was satisfied with being in charge of a relatively small institute where he could take pride in knowing about all its operations. He did not yearn for official positions except for the directorship of the institute he had founded and for being in charge of the oxygen authority, responsible for the oxygen production, which was crucial especially during the war. He liked meditation and he liked to exert his influence from the background. His lack of desire for publicly visible power made it possible for him to maintain his principles and act upon them that other Soviet scientists of similar stature found impossible to accomplish.

Yuly Khariton [9/2] (Chapter 1).

Yuly Khariton's grave contains not only his remains, but also his wife's (Mariya N.), his daughter's (Tatyana Yu.), his son-in-law's (Yury N. Semenov) and his son-in-law's mother's (Natalya N. Semenova). The latter was Nikolai Semenov's divorced second wife. According to a private communication from his grandson, Aleksei Semenov, when Khariton died, the family asked their friend, the sculptor Nikolai Silis for his assistance in choosing an appropriate memorial. Silis suggested the abstract sculpture that now stands next to the larger stone with the inscription. It is a piece of red granite with two asymmetrically positioned holes as if representing a heart.

Aleksandr A. Kharkevich (1904–1965) [3] was a theoretical physicist of what is now the Kharkevich Institute for Information Transmission Problems (19 Karetny Passage).

Lev N. Khitrin (1907–1965) [6/14] researched thermal physics, thermal energy, combustion, and chemical physics at what is now the Krzhizhanovsky Energy Institute (19 Leninsky Avenue). He was also was professor and head of the department of combustion at the Faculty of Physics of Moscow University.

Rem Khokhlov [7/13] (Chapter 2).

Isaak Kikoin [10/2] (Chapter 1).

Lev Landau [5/38] (Chapter 1).

Lev Landau's gravestone is the creation of Ernst Neizvestny. The bust rests on a tall metallic stand. Neizvestny was a persona non grata in the Soviet system (see above), but the top physicists supported him with commissions and with exhibiting his art at their institutes. Kapitsa insisted that Neizvestny should be commissioned to create Landau's tombstone. Landau's longtime collaborator, academician Evgeny M. Lifshits (Chapter 1), is buried at the Kuntsevskoe Cemetery (Chapter 8).

Grigory S. Landsberg (1890–1957) [5/1] was a graduate and a professor of physics of Moscow University and he also taught at the Moscow Institute of Physics and Technology. Together with Leonid Mandelshtam he discovered the phenomenon of combination scattering of light (see below in the Mandelshtam entry).

Petr P. Lazarev [1/8] and Petr Lebedev [3/20] (Chapter 1).

Petr P. Lazarev (1878–1942) founded and edited for many years the Soviet physics journal that appears in English as *Physics — Uspekhi: Advances in Physical Sciences*. Lazarev was a disciple of Petr Lebedev and when Lebedev died, Lazarev took over Lebedev's former laboratory. After the 1917 revolution, Lazarev took up the responsibilities for medical instrumentation related to X-rays, electromagnetic apparatus, and photo-biological instrumentation in the ministry of health. In 1919, Lazarev initiated the first research institute of physics after the revolution and he induced many of the best physicists in Russia to join this institute.

Lazarev was an independent-minded scientist and thus an anathema to the new powers that be. In 1929, he criticized that communists were favored in the elections to the Academy of Science; he pointed out a factual error in Friedrich Engels's works; and his wide-scale correspondence was also a thorn in the side of the authorities. In 1931, Lazarev was arrested; he was stripped of his positions; his associates were fired; and his experimental apparatus disappeared. In this hopeless situation, Lazarev's wife committed suicide. Lazarev was exiled, but could return to Moscow in 1932; by then, he was a sick man. Scientists, former associates and former pupils, had been asking the authorities to free Lazarev. Some though did not sign such requests, for example, the Vavilov brothers, probably out of fear of repercussions. From 1934, Lazarev was head of the section of biophysics at the Institute of Experimental Medicine.

Leonid I. Mandelshtam (1879–1944) [4/40] and Grigory Landsberg (see above) discovered what in Russia is known as the effect of *combination scattering* and in the rest of the world as the Raman Effect. Its essence is that when light is shone onto a molecule, the molecule deflects a small portion of it with frequencies that are a combination of the frequency of the incoming light and the frequencies of molecular vibrations. Hence is the meaning of the name of combination scattering. This is how information about the vibrations of the molecule may be extracted from the observation of the deflected light. Although in 1927–1928, Mandelshtam and Landsberg conducted their experiments shortly before the Indian Chandrasekhara Venkata Raman, the Indian scientist published his independent observations before the Russian scientists did. Raman received the Nobel Prize in Physics in 1930 for his work on light scattering and for the discovery of the Raman Effect. In hindsight, the Nobel Prize would have been justified to divide one half to Raman and the other half to Mandelshtam and Landsberg. Mandelshtam made other significant contributions to physics and built up an influential school. Igor Tamm revered him as his teacher and Vitaly Ginzburg considered himself a member of the Mandelshtam-Tamm School.

Arkady B. Mikdal (1911–1991) was a theoretical physicist who worked in various research institutes of the Academy of Sciences and taught at the Moscow Institute of Physical Engineering. Many of his former pupils became internationally renowned physicists. From 1945, he participated in the Soviet nuclear project. He died in Princeton during a visit there.

Pavel P. Parenago (1906–1960) [5/30] was a graduate and subsequently a professor of Moscow University. He founded and led the department of stellar astronomy, the branch of astronomy studying the properties and evolution of stellar systems.

Aleksandr Prokhorov [11/5] (Chapter 1).

Dmitry Skobeltsyn [10/6] (Chapter 2).

Igor Tamm [7/14] (Chapter 1).

Igor Tamm has an unusual tombstone by the sculptor Vadim Sidur (Chapter 8), especially if considering the classical style of tombstones in its surroundings. Vitaly Ginzburg took Tamm for a visit to Sidur's studio and Tamm had mixed feelings about what he saw. When Tamm died, the authorities wanted a classical style tombstone, but Tamm's family insisted on asking Sidur. The family felt that although Tamm might have had a classical taste in art; he had a maverick personality and mobile demeanor and Sidur would better correspond to his nature. Apparently Sidur was inspired to create a masterpiece.

Sergei Vavilov [1/42] (Chapter 1).

Vladimir I. Veksler [6/34] (Chapter 1).

Sergei Vernov [10/1] and Yakov Zeldovich [10/4] (Chapter 1).

Sergei N. Vernov (1910–1982) started his career at FIAN, but in 1943, he moved to the Faculty of Physics of Moscow University. He participated in establishing a separate research institute of nuclear physics at the Faculty of Physics, served as deputy director, and from 1960, its director until the end of his life. High-energy physics was his field of interest and he created the experimental conditions for detecting extremely high-energy cosmic rays. His research of cosmic rays assisted the space program.

Bentsion M. Vul [10/3] (Chapter 1).

Rocket and Space Scientists and Technologists

Georgy N. Babakin (1914–1971) [7/17] had a leading role in designing spaceships.

Vladimir Barmin [10/7] (Chapter 4).

Anatoly Blagonravov [6/28] (Chapter 1).

Vladimir Chelomei [7/] (Chapter 4).

Boris Chertok [11/6].

Boris E. Chertok (1912–2011) spent sixty years in the development of aviation and in space exploration. He published his memoirs in a book series *Rockets and People*, which has appeared in English translation in the framework of the NASA History Series. Chertok's tombstone depicts his four volumes with one of them opened. The loose translation of the text in the open pages is as follows: "I count myself to be a member of the generation that suffered irreplaceable losses for which the heaviest ordeals followed in the 20th century. From their childhood, the feeling of duty was inculcated in this generation — duty to the fatherland, to later generations, and even to the whole of mankind."

Valentin Glusko [10/5] (Chapter 4).

Petr Grushin [10/8] and Semen Lavochkin [1/3].

Petr D. Grushin (1906–1993) constructed airplanes and rockets. He was Semen Lavochkin's deputy during WWII. After the war, he worked in jet technology for the Ministry of Aviation Industry. He focused his attention to the development of rocket technology. He is credited to be one of the architects of the technology that brought down the U-2 American reconnaissance plane on May 1, 1960, causing an international scandal.

Aleksei M. Isaev [4/] (Chapter 4).

Georgy F. Katkov (1919–1989) [7/3] was a designer whose activities concerned the electrical apparatus of rockets and vehicles in cosmonautics.

Viktor I. Kuznetsov [11/2] (Chapter 4).

Semen A. Lavochkin (1900–1960) participated in the civil war before he was directed to study at Bauman University. He graduated as engineer majoring in aero-mechanics. His designs demonstrated high quality in WWII. After the war he worked on jets and from the mid-1950s, on the development of intercontinental supersonic rockets.

Viktor P. Makeev (1924–1985) [10/3] dealt with the development of strategic rockets for the Soviet Navy and founded a school for training specialists in this area.

Aleksandr Nadiradze [10/4] and Ari Sternfeld [5/20].

Aleksandr D. Nadiradze (1914–1987) was a rocket designer who applied the latest achievements of mechanics in rocket development.

Boris N. Petrov (1913–1980) [9/6] was a researcher of automated regulation and worked in close cooperation with the leading designer of the space program, Sergei Korolev.

Nikolai Pilyugin [10/1] (Chapter 4).

Ari A. Sternfeld (1905–1980) was born in Poland into a Jewish family. He studied in Krakow, Nancy, and Paris at the Sorbonne. From early youth he was addicted to learning about the Universe, coined new names, corresponded with Tsiolkovsky, and published numerous articles and books about cosmonautics. He lived in Paris, but in 1932 he visited the Soviet Union and he and his wife settled there from 1935. He began his Soviet career in a research institute for jet-propelled aviation, in the section led by Korolev;

Glushko and Mikhail Tikhonravov were among his colleagues. During Stalin's terror in the late 1930s, the director of the institute, I. T. Kleimenov, and the chief engineer, G. E. Langemak, were executed; Korolev, Glushko, and others were exiled; Sternfeld was "only" kicked out of the institute. For the rest of his life he was never to find employment again although he appealed to Stalin, Fesenkov, Komarov, Keldysh, Sergei Vavilov and others for help to find a job; there was never a response. Sternfeld continued his work at home, by himself, supported by his wife and, eventually, by his writing. From the late 1950s, his writings about satellites and space research attracted international interest. This interest was enhanced by the Soviet success in space research. From the early 1960s, Sternfeld started receiving recognition at home as well. His tombstone (by F. S. Khazan) labels him "Pioneer of Cosmonautics." It quotes the ancient saying, "per aspera ad astra ("through barriers to the stars"). There is a trajectory of a satellite carved into his tombstone, which appears as if depicting an open book.

Mikhail K. Tikhonravov (1900–1974) [7/8] was an engineer who designed rockets and from early on worked with Sergei Korolev.

Leonid A. Voskresensky (1913–1965) [6/29] was one of Korolev's close associates. His principal work was in testing the components of rocket technology.

Mikhail K. Yangel (1911–1971) [7/1] was a much decorated designer of rocket systems and systems for space exploration.

Airplane Designers

Aleksandr A. Arkhangelsky (1892–1978) [1/] was a high ranking airplane constructor. He studied under Nikolai Zhukovsky and worked with Andrei Tupolev, including the time they shared as slave laborers in "sharashkas."

Rostislav Belyakov [1/41].

Rostislav A. Belyakov (1919–2014) had the rank of general designer during the last period of his career. His principal interest was in designing small airplanes capable of operating by automata. He designed large airplanes as well that had the capability of operating under extreme physical conditions. The grave contains his wife's and son's remains beside his own. The tombstone is the creation of Mikael Sogoyan.

Aleksei Cheremukhin and Georgy Cheremukhin [5/24].

Aleksei M. Cheremukhin (1895–1958) constructed the first Soviet helicopter and test flew it to such a height (605 meter) that was the unofficial record of the time. In 1938, he was arrested and spent three years as slave laborer in a "sharashka" where he held responsible positions. He was rehabilitated only in 1955. Next to Aleksei Cheremukhin's tombstone is the tombstone of his son, Georgy, displaying a supersonic airplane. Georgy A. Cheremukhin (1921–2009), just as his father, spent his entire career designing and improving airplanes.

Sergei Ilyushin [7/13] and Gleb Kotelnikov [7/1].

The aircraft designer Sergei V. Ilyushin (1894–1977) had a high military rank, and he was also a member of the Academy of Sciences. His name was well known even for people not involved with aircraft technology. He designed the airplane with which an internationally recognized world record was established — it lifted a record weight to a record height in 1936. In WWII, Ilyushin's IL-4 bomber and IL-2 fighter became legendary. Soviet flyers flew their IL-4s to Berlin even at the time of the worst retreat of the Soviet military during the fateful August and September of 1941. The war was not yet over when Ilyushin was charged with designing civilian aircraft and the results included IL-12, IL-14, IL-18, and the flagship of Aeroflot in the 1960s and 1970s, IL-62.

Nikolai I. Kamov (1902–1973) [7/7] was a pioneer in producing helicopters in the Soviet Union.

Vladimir Ya. Klimov (1892–1962) [9/4] designed airplane engines. Parallel to his career in academia, he was an air force general.

Semen A. Kosberg (1903–1965) [6/14] was a specialist of air plane and rocket engines. He graduated from the Leningrad Polytechnic Institute and worked in Moscow in what is today the Baranov Central Institute of Aviation Motors (2 Aviamotornaya Street).

Gleb E. Kotelnikov (1872–1944) invented a parachute. He attended a military school, but he soon switched to civilian occupations and used his spare time for his work on his invention. The origin of his dedication to inventing a parachute was the senseless death of a flyer. He submitted his first patents in 1912 in France, because the previous year he did not succeed in getting his patent recognized in Russia. He continued improving his parachute in Soviet times. Kotelnikov's impressive tombstone has become a place of an annual pilgrimage for parachutists. They have developed a habit of tying parachute cords to the trees around Kotelnikov's grave as if to register their presence.

Viktor S. Kulebakin (1891–1970) [7/12] was an air force general and an electrical engineer. He designed airplanes.

Arkhip M. Lyulka (1908–1984) [7/20] designed a gas-turbine transmission employed in a number of airplanes associated with renowned engineers such as Pavel Sukhoi, Sergei Ilyushin, and Andrei Tupolev.

Artem I. Mikoyan (1905–1970) [1/42] was an engineer and involved in quality control of airplane designs.

From left to right: Nikolai Polikarpov [1/43], Pavel O. Sukhoi [7/11] (Chapter 4), and Sergei Tumansky [7/6].

Nikolai N. Polikarpov (1892–1944) was one of the founders of airplane design in the Soviet Union. He studied at the St. Petersburg Polytechnic Institute and began his activities by improving already existing aircraft. His promising career was interrupted by his arrest in 1929. He was accused of counter-revolutionary activities and sabotage. A number of other designers in the aircraft industry were also arrested at the same time under similar charges. He was among those sentenced to death without trial, but he was not executed, as one of the exceptions. A design bureau was organized to employ the incarcerated engineers. Polikarpov was appointed deputy head of this bureau.

In 1930, the first fighter plane from this bureau tested successfully. In 1931, a team led by Polikarpov, was charged with designing a new fighter plane. A little later Polikarpov was sentenced to ten years of slave labor. Still in 1931, the fighter planes designed by Polikarpov and his team of slave laborers were demonstrated to Stalin and his comrades with great success. Polikarpov was freed, but his rehabilitation came only in 1956, 12 years after his death. In 1935, he received a high government award followed by further distinctions over the years. He continued his work, designed airplanes that were deployed by the air force with great success and he was appointed to ever higher positions. He was even sent for a visit to Germany in 1939. When he died, Vladimir Chelomei was appointed to be director of the aviation factory in his stead (Chapter 4).

Nikolai K. Skrzhinsky (1904–1957) [5/14], a graduate of the Kiev Polytechnic Institute, co-designed the first Soviet rotary-wing aircraft.

Sergei K. Tumansky (1901–1973) designed jet engines and other aircraft engines and researched the ideal conditions for operating turbines at high temperatures. He was killed in a work-related accident.

Andrei Tupolev [8/46] (Chapter 4) and Aleksandr Yakovlev [11/1].

Aleksandr S. Yakovlev (1906–1989) was already building planes as a teenager. He flew gliders and created one in 1924 that was judged to be the best Soviet glider. He wanted to study to become a flyer, but could not enroll because of his non-proletarian origins. The years 1924–1927 were full of frustration for him until he was accepted in 1927 to become a student of the Zhukovsky Air Force Academy. He graduated in 1931 and soon organized a group to design light planes. From this point on his career was successful, even too successful because at some point he was appointed deputy minister for the airplane industry, which took him away from his beloved occupation. After a few years he resigned and returned to designing aircraft. His Yak airplanes established 74 world records. Yakovlev's bust (M. K. Anikushin) stands in the Park of Aviators.

Engineers, Technologists, and Inventors

Pavel V. Abramov (1902–1956) [1/41] was a civil engineer — an architect. He had a leading position in Glavmosstroi — today, this is a big company for building homes.

Pavel V. Abrosimov (1900–1961) [8/10] architect was engaged in building large complexes in Leningrad, Moscow, Kiev, and elsewhere. He was one of the builders of the high-rise of the new Moscow University.

Karo S. Alabyan (1897–1959) [5/31] was an Armenian architect who started in Erevan, then moved to Moscow and was having a brilliant career as a builder and as an official in the organizations of architects. However, he had a disagreement with the all-powerful Beriya and found himself out of favor. He was saved by returning to Armenia and getting out of the sight of the secret police.

Aleksandr Aleksandrov [9/7] and Ivan Bardin [1/41] (Chapter 1).

Aleksandr P. Aleksandrov (1906–1981) [9/7] worked as an engineer architect for the security organs on projects of buildings and highways using slave labor. He took part in creating the navigable canal connecting the rivers Volga and Don and in building hydroelectric power plants. Such a structure is depicted on his gravestone.

Ivan Artobolevsky (1905–1977) (Chapter 1).

Vladimir I. Dikushin, (1902–1979) [9/5] engineer perfected how metals are cut and improved machine tools.

Vladimir Kirillin (Chapter 4).

Leonid S. Leibenzon (1879–1951) [3] was a disciple of Nikolai Zhukovsky. Leibenzon graduated from Moscow University and from what is today Bauman University. He had a professorial appointment in applied mathematics at Moscow University, which he left in protest in 1911. He worked in the oil industry under Vladimir Shukhov. He was back at Moscow University from 1922. He was recognized for his contributions to the oil industry and to aviation. He was incarcerated between 1936 and 1939 and exiled together with his pediatrician wife. Sergei Chelomei (Chapter 4) and others took a stand to have him freed. Leibenzon then worked at Moscow University until the end of his life.

Ivan K. Matrosov (1886–1965) [6/25] was an inventor of automatic railroad breaks.

Nikolai P. Melnikov (1908–1982) [10/1] specialized in the mechanics of architecture and especially in metallic structures.

Nikolai I. Mertsalov (Chapter 4).

Mikhail A. Mikheev (1902–1970) [7/15] researched thermal power plants and was concerned with energy production and economy.

Mikhail Millionshchikov [7/4] (Chapter 1).

Aleksandr L. Mints [7/10] (Chapter 1).

Vasily P. Nikitin (1893–1956) [4/18] specialized in welding and other areas of electrical engineering.

Nikolai Nikitin [7/4].

Nikolai V. Nikitin (1907–1973) was Vasily Nikitin's son. He was an architect and participated in creating such famous structures as the new Moscow University, the Ostankino TV tower, and the central sports arena in Luzhniki. His tombstone displays a stone carving of the Ostankino TV tower.

Valery I. Popkov (1908–1984) [10/2] was much involved in the development of electrical engineering, the training of electrical engineers, and in energy production and economy.

Valentin L. Pozdyunin, (1883–1948) [3/47] was a shipbuilder engineer and his research concerned hydromechanics. His inventions helped to perfect naval operations.

Aleksandr Raspletin [6/27] and Kirill Shchelkin, [6/2)].

Aleksandr A. Raspletin (1908–1967) constructed devices in radio engineering and electronics. His innovations found applications in multi-head rocketry, in the modernization of anti-aircraft defense, and in the development of anti-rocket defense.

Kirill I. Shchelkin (1911–1968) was the head of the second Soviet nuclear laboratory, called Chelyabinsk-70, in Snezhinsk in the Ural region. He was one of the most decorated Soviet engineers. His own research was in combustion and detonation and in turbulence. He moved from Arzamas-16 to Chelyabinsk-70 when it started its operations in 1957. When he returned to Moscow, he was appointed to head the department of combustion at the Moscow Institute of Physics and Technology. He lectured on popular science and published a well-received series, *Physics of the Microworld.*

Vladimir Shukhov [2/40] (Chapter 4) and Pavel Zernov [6/4].

Nikolai S. Streletsky [6/37] (Chapter 8).

Nikolai D. Ustinov (1931–1992) [6/28] was a radio physicist and engineer; he specialized in laser technology.

Aleksandr V. Vinter (1878–1958) [5/18] built and operated electrical power plants.

Boris A. Vvedensky (1893–1969) [7/8] was a radio physicist who co-founded the Kotelnikov Research Institute of Radio Physics and Electronics. He investigated the nature and transmission of radio waves and analyzed the connection between the ultra-shortwave transmission and the meteorological conditions of the atmosphere.

Pavel M. Zernov (1905–1964) was a high-ranking military officer of the engineering service. Between 1951 and 1964 he was a deputy minister in the Ministry of Medium Machine Building, which was in charge of the nuclear weapons program. Zernov's role might be compared to the task of General Leslie Groves, the military commander in the Manhattan Project. Zernov had shown talent in directing various industrial projects in the prewar period as well as in war-time. In 1946, the Soviet leadership appointed Zernov to be the military commander of the development of nuclear weapons. It was under his supervision that the collective of scientists and technologists was brought together at Arzamas-16. The necessary infrastructure was built by slave labor. Upon the successful test of the atomic device in 1949, Zernov received the highest accolades along with the leading scientists.

Chemists and Materials Scientists

Kuzma Andrianov [9/3] (Chapter 1), Aleksei Bakh [4/24] (Chapter 1) and Aleksei Balandin [6/34].

Aleksandr Baikov [2/4] (Chapter 1).

Aleksei A. Balandin (1898–1967) studied physical chemistry at Moscow University and from 1934, he was professor of physical chemistry there. On trumped up charges, he was exiled between 1936 and 1939 and worked in an analytical chemistry laboratory. Courageous scientists, such as Zelinsky, Bakh, Frumkin, Vernadsky, Kurnakov, and others, declared that Balandin could not have conducted anti-Soviet activities. He was freed and upon his return, he organized a new laboratory at the University for organic catalysis. Catalysts are "chemical matchmakers" making it easier for the reactants to enter a reaction. During the war, Balandin worked on defense-related projects. In 1948, he was appointed Dean of the Faculty of Chemistry.

Balandin was arrested again in 1949 together with a group of scientists, mostly geologists, falsely accused of sabotaging uranium exploration. Four professors did not survive the ordeal, the rest, close to 200 people, received sentences ranging from 10 to 25 years. During Balandin's incarceration, it happened that at a banquet following the successful defense of a dissertation at Moscow University, a student proposed a toast for the health of Academician Balandin. A commission investigated the "crime," the student was expelled from Moscow University, and his mentor was fired.

Balandin was freed in 1953 and appointed to be in charge of the department of organic catalysis at the University along with other positions in academia. He was an internationally renowned scientist, especially for his original theories about the mechanism of catalysis. At the base of his tombstone, molecular models representing his theory are carved into the stone, but the carving has faded and is hardly visible.

Edgar V. Britske (1877–1953) [1/9] was a chemist by training, but made his career in metallurgy. His interest included the utilization of chemical fertilizers and defense against chemical weapons.

Petr P. Budnikov (1885–1968) [7/6] graduated in chemistry and started his career in a chemical plant in what is today Poland. He visited England and France on assignment from the czarist ministry of education. After the revolution he became internationally renowned for his research in inorganic materials, especially silicates and building materials. He remained involved in industrial projects, but during his last decades his main interest was his professorial activities at the Mendeleev University of Chemical Technology.

Nikolai P. Chizhevsky (1873–1952) [1/14] studied at St. Petersburg University but one month before graduation he was excluded from studying in all institutions of higher education in Russia for having participated in an illegal protest demonstration. He continued his studies in Austria, and started his career at the Kiev Polytechnic Institute. Then, he taught at the Tomsk

Polytechnic Institute. Finally, he moved to Moscow. He worked at the Mining Academy and in the Institute of Steel, and ultimately at the Petroleum Institute of the Academy of Sciences. He introduced innovations in the production of coke, iron, steel, and graphite.

Nikolai Emanuel [10/2] (Chapter 1).

Nikolai Enikolopov [10/7] (Chapter 1).

Aleksandr Frumkin [4/53], and Vitaly Goldansky [10/5] (both in Chapter 1).

Aleksandr Frumkin's second wife, Amaliya D. Obrucheva, was his close associate. After her death, Frumkin married Emiliya G. Perevalova (1922–2012), an organic chemistry professor at Moscow University. The remains of Frumkin and Perevalova rest in their joint grave. She commissioned the tombstone after Frumkin's death. When she visited Vadim Sidur's studio, she noticed two figures cuddling up to each other expressing sorrow and grief, but not despair, and she knew she found what she was looking for. The tombstone has become popular among Frumkin's former students and associates and they gather at it annually on the day of the anniversary of Frumkin's death to remember their teacher and revered colleague.

Nikolai T. Gudtsov (1885–1957) [5/1] was an engineer who gained considerable industrial experience before he joined academic research in metallurgy. During WWII he was engaged with the production of steel of the quality that would satisfy the demands of weapons production.

Vladimir Gulevich [3/34] (Chapter 2).

Viktor Kabanov [11/5] (Chapter 2).

Sergei Kaftanov [9/3] (Chapter 4).

Anatoly F. Kapustinsky (1906–1960) [8/5] was a physical chemist who taught at Moscow University and elsewhere. He wrote about the history of inorganic and physical chemistry in Russia and about famous chemists.

Valentin Kargin [7/9] (Chapter 2).

Boris A. Kazansky (1891–1973) [7/5] worked in the chemistry of hydrocarbons — compounds consisting of hydrogen and carbon. He was one of the founders of petroleum chemistry and catalysis.

Bonifaty M. Kedrov (1903–1985) [10/3], a chemist by training, was a leading Soviet philosopher of natural sciences. He was active in the ideological interference in chemistry in the late 1940s and early 1950s.

Vladimir Kistyakovsky [4/12] (Chapter 1).

Isaak Kitaigorodsky and his son, Aleksandr Kitaigorodsky [6].

Isaak I. Kitaigorodsky (1888–1965) graduated from the Kiev Polytechnic Institute and specialized in the chemistry and technology of glass and ceramics. He co-owned a glass-ceramics plant before the 1917 revolutions, which ended his proprietorship, but he was retained as director of the plant in recognition of his expertise. He was even sent to Western Europe and the United States to broaden his experience. In 1929, he was given the title of professor and when the new system of scientific degrees was introduced in 1934, he received the degree of Doctor of Science. His products were valued nationwide and when a special glass was needed for a frontier experiment in physics, they turned for it to Isaak Kitaigorodsky.

At the foot of Isaak Kitaigorodsky's tombstone there is a black plate indicating the grave of his son, Aleksandr Kitaigorodsky (Chapter 1). He was a great scientist and original thinker who built up a unique career in crystallography and his former pupils have spread his teachings worldwide. He predicted the relative frequencies of various symmetries expected among crystal structures long before it could be deduced on the basis of experimental data. He had a deep understanding of nature and the vision to make such predictions. Aleksandr Kitaigorodsky was a respected scientist with the title of professor and in charge of a laboratory at the Institute of Element-organic Compounds. However, he could never have a chair at Moscow University and was never elected to the Academy of Sciences. He was flamboyant and irreverent toward the authorities of the scientific establishment, and another significant reason was anti-Semitism. This was not a factor in Soviet scientific life when Isaak Kitaigorodsky embarked on his career in the young Soviet state. It became an important factor after the war, during Stalin's last years and to a considerable degree for the rest of the existence of the Soviet Union.

Pavel I. Korobov (1902–1965) [6/21], engineer metallurgist, occupied important industrial and governmental positions.

Boris Kurchatov [7/1] (Chapter 1).

Valery Legasov [10/5] (Chapter 1) and Aleksandr Nesmeyanov [9/5] (Chapter 1).

Sergei S. Medvedev [7/15] (Chapter 4).

Sergei S. Nametkin (1876–1950) [3/16] graduated from Moscow University where he was Nikolai Zelinsky's student. Nametkin specialized in petroleum chemistry. He taught at Moscow University, and left it in protest against the arbitrary policy of the tsarist minister of education. Nametkin taught at the Higher Courses for Women and continued there after the revolution when the Courses were transformed into the Second Moscow State University. Nametkin focused his attention on research and had professorial appointments at the Gubkin Institute and at the Lomonosov Institute of Fine Chemical Technology, as well as a leading research position at what is today the Topchiev Institute of Petroleum Chemical Synthesis.

Yury A. Ovchinnikov [7/2] (Chapter 1).

Sergei S. Perov, (1889–1967) [7/2] investigated the biochemistry of proteins, especially the proteins of milk. He was a supporter of Trofim Lysenko.

Nikolai A. Plate (1934–2007) [4/41] was another of Valentin Kargin's pupils (Chapter 2) in polymer science. Plate excelled in contributing to the development of polymers that could be used for a variety of functions in clinical practice and of biocompatible materials for the preparation of prostheses.

Vladimir M. Rodionov (1878–1954) [3/60] distinguished himself by the broad range of his organic chemistry and his interest in assisting the industrial production of dyes, pharmaceuticals, and alkaloids.

Boris A. Sakharov (1914–1973) [7/5] was a physical chemist and chemical technologist. He worked for producing new semiconductors and metals of ultrahigh purity. His gravestone is decorated with symbols of semiconductors.

Aleksandr M. Samarin (1902–1970) [7/13] researched the interaction between gases and steel in order to improve steel production.

Nikolai P. Sazhin, (1897–1969) [6/22] was a metallurgist specializing in rare metals and materials of high purity.

Nikolai N. Semenov [10/4] (Chapter 1), Mikhail M. Shemyakin (Chapter 1), and Nikolai Zelinsky [1/42] (Chapter 1).

According to family members, Nikolai N. Semenov's statue on his grave shows him too pretentious whereas he was an informal, friendly, even earthy person. Semenov's first wife, Maria Boreisha, was seventeen years his senior, had already been married with children, when they met. She was vivacious, always the center of society, but succumbed to cancer. Semenov's second wife was Maria's niece, Maria Burtseva. They had two children, Yury who married Yuly Khariton's daughter and Lyudmila who married Vitaly Goldansky. Semenov was 75 years old when he divorced his second wife and married his third, one of his assistants at his Institute. All his old friends, including the Kapitsas, remained friends with his former second wife. She and their son Yury are buried in Khariton's grave. Lyudmila is still alive, but she will eventually join Goldansky.

Boris P. Terentiev (1899–1995) [4/36] as organic chemist synthesized new substances and perfected the production of organic compounds for industry.

Aleksandr Topchiev [8/23] (Chapter 4).

Georgy G. Urazov (1884–1957) [5/6] was a specialist of chemical technologies in the production of metals, including aluminum. From 1943, he was in charge of the division of chemical analysis at the Kurnakov Institute of Inorganic Chemistry.

Vladislav Voevodsky [6/37] (Chapter 2).

Anton N. Volsky, (1897–1966) [6/29] worked in metallurgy and chemical technology. He pioneered the utilization of information about chemical equilibria in melts of metals, and contributed to the theory of metallurgical processes.

Boris P. Zhukov (1912–2000) [10/6] was an industrial chemist. He was the leader of the group of scientists and engineers that created the best chemical composition for preparing the charges of the "Katyushas" during WWII.

Biologists and Agricultural Scientists

Vasily V. Alekhin (1882–1946) founded the department of geo-botany at Moscow University. He also taught at the Higher Courses for Women. He did fundamental work in the morphology of plants and in creating the botanical geography of the Soviet Union.

Boris Astaurov [7/1] (Chapter 1).

Sergei F. Averyanov [4/38] (Chapter 6).

Andrei Belozersky [7/13] (Chapter 2).

Mikhail I. Dyakov (1878–1952) [4/21] was a zoo-technologist and professor at several institutions of higher agricultural education.

Vitaly Edelshtein [6/21] (Chapter 6).

Left: Vladimir Engelhardt [10/2] (Chapter 1). *Right*: Merkury Gilyarov [10/3] (Chapter 1). Gilyarov's gravestone hints at his love of the animal world.

Leonid A. Ivanov (1871–1962) [4/39] was a botanist with his principal interest in plant physiology. He worked mostly in research institutes of the Academy of Sciences.

Vladimir Komarov [1/30] (Chapter 1).

Andrei L. Kursanov (1902–1999) [3/9] was a renowned scientist of plant physiology. The Institute of Biochemistry, Moscow University, and the Timiryazev Academy were among the venues where he worked as a scientist and educator. His father, Lev. I. (1877–1954), and uncle, Nikolai I. (1874–1921), were also famous biologists and rest in the same grave.

Left: Efim Liskun [5/19] (Chapter 6). *Right*: Sergei Muromtsev (bust on the left) and Georgy Muromtsev (bust on the right) [8/6].

Sergei N. Muromtsev (1898–1960) and his son, Georgy S. Muromtsev (1932–1999) worked in agricultural microbiology. The father specialized in medicinal and veterinarian microbiology and the son was more involved with biotechnology.

Aleksandr Oparin [9/6] (Chapter 1).

Yury A. Orlov [6/27] (Chapter 8).

Aleksandr S. Serebrovsky (1892–1948) [4/15] graduated from Moscow University and fought in WWI before he started his career as a geneticist. He worked at Koltsov's Institute of Experimental Biology and at the Timiryazev Institute of Biology. From 1930 until the end of his life, Serebrovsky was in charge of the department of genetics of the Faculty of Biology at Moscow University. He appeared immune to Lysenkoism, and condemned the slogans counterpoising "true Soviet genetics" and "bourgeois genetics."

Vladimir N. Shaposhnikov (1884–1968) [4/49] graduated from Moscow University and started his career at its department of plant physiology and anatomy under Timiryazev's mentorship. Shaposhnikov was a professor at Moscow University and worked also for the Academy of Sciences. He founded technical microbiology in the Soviet Union.

Norair Sisakyan [6/31] and Konstantin Skryabin [7/2] (Chapters 1 and 5).

Norair M. Sisakyan (1907–1966) researched the mechanism of fermentation, the processes of metabolism, the biochemistry of drought-resistant plants, technical biochemistry, and the biology of Space. He worked at the Bakh Institute of Biochemistry and taught at Moscow University. Sisakyan was a supporter of Lysenko; he represented the Soviet Union in a number of international science organizations, where he was elevated to high positions, and assisted the formation of false reality about Soviet science internationally.

Georgy K. Skryabin (1917–1989) [10/5], Konstantin Skryabin's (Chapter 1) son, was a microbiologist and for a long time the powerful permanent secretary of the Academy of Sciences.

Lina Stern [6/36] and Nikolai Tsitsin [9/6] (Chapter 8).

Lina S. Stern (1878–1968) originated from the western part of today's Latvia. She studied biochemistry and neurology in Geneva and became the first female professor at the University of Geneva. In 1925, she moved to the Soviet Union at the invitation of the Soviet government. She uncovered what is called today the blood-brain barrier, which is a barrier of selective permeability between the circulating blood and the extracellular fluid of the brain. She became the first female full member of the Soviet Academy of Sciences — an academician. During the war her discoveries turned into life-saving approaches in the treatment of heavily wounded soldiers. She was a member of the Jewish Antifascist Committee and of the Soviet Women's Antifascist Committee. After the war, Stern suffered from Stalin's anti-Semitic policies and from Stalin's anti-science crusades. She was stripped of all her positions, incarcerated, and exiled far from Moscow. She was the most fortunate member of the Jewish Antifascist Committee, because all the others were executed. She returned to Moscow after Stalin's death; she was exonerated, and her membership in the Academy of Sciences was reinstated. She resumed her research activities as head of the department of physiology at the Institute of Biophysics.

Elizaveta I. Ushakova (1895–1967) [8/44] became the first woman member of the Academy of Agricultural Sciences in 1948. As a vocal supporter of Lysenko, she condemned in the strongest terms the "idealist" and "reactionary" Soviet followers of Mendel and Morgan.

Vadim M. Yudin (1899–1970) [7/15] was a zoo-technologist and head of departments of animal husbandry and veterinary sciences at institutions of agricultural education.

Lev A. Zilber, Lev L. Kiselev, and Fedor L. Kiselev [6/25] (Chapters 1, 5, and 8).

Medical Scientists and Physicians

Left: Aleksei Abrikosov and Fanni D. Abrikosova (née Vulf) [3/42] (Chapter 5). *Right*: Aleksandr Bakulev [6/35].

Aleksei I. Abrikosov and Fanni Abrikosova's son, the future Nobel laureate physicist, Aleksei A. Abrikosov, was Lev Landau's pupil at the Institute of Physical Problems. Following graduation, Aleksei would have liked to continue working with Landau. This was in the early 1950s, and the security organs prevented this, supposing that Aleksei, having a Jewish mother, would be a security risk; so he had to find work somewhere else. It happened that the Mongolian communist leader, Khorloogiin Choibalsan, died in Moscow and the communist party newspaper *Pravda* published the medical report of his death, which was signed, among others, by Aleksei's pathologist mother. The security people were duly impressed that she had been entrusted this most important autopsy, and let Aleksei back to the Institute of Physical Problems.

Petr K. Anokhin [7/left side6] (Chapter 5).

Andrei A. Bagdasarov (1897–1961) [8/7] was a physician who participated in developing blood conservation and transfusion. He was also successful in immuno-hematology and worked as the chief hematologist of the Ministry of Health.

Aleksandr N. Bakulev (1890–1967) specialized in the surgery of the cardiovascular system. There is today a Bakulev Research Center of Cardiovascular Surgery (8 Leninsky Prospekt). The tombstone on Bakulev's grave symbolizes the surgeon's two hands protecting the human heart.

Nikolai Blokhin [10/7] (Chapter 5).

Nikolai Burdenko [1/42, bust by G. Postnikov] (Chapter 5).

Ippolit Davydovsky [7/right side3] (Chapter 5).

Left: Boris G. Egorov and his cosmonaut physician son Boris B. Egorov [7/16] (Chapter 5). *Right*: Nikolai Elansky [6/10] (Chapter 5).

Vladimir D. Fedorov (1933–2010) [4/45] was an innovative surgeon and director of the Vishnevsky Institute of Surgery. His wife, Maina V. Fedorova (1932–2008), buried in the same grave, was also a physician, a professor of obstetrics.

Left: Nikolai Gamaleya [4/39] (Chapter 5). *Right*: Aleksei Ochkin [2/28] (Chapter 5); the bust on Ochkin's gravestone is by Z. I. Azgur.

Petr B. Gannushkin (1875–1933) was a psychiatrist who founded a school of psychiatry and was a professor at Moscow University. In 1911, along with other renowned professors, he left the University in protest. Gannushkin worked in a hospital and for the first war years, he served in the Russian army. In 1917, he returned to his work in the hospital and after the revolutions he resumed his professorship at Moscow University.

Petr A. Gertsen (1871–1947) [4/41] originated from a family of great intellectuals. His grandfather was the famous Russian writer and thinker, Alexander I. Herzen and his father, the renowned Swiss physiologist Aleksandr A. Herzen. Gertsen was a surgeon, did research in a variety of the medical professions, and a whole series of procedures and approaches have been named after him. His contributions enriched the surgery of the abdomen, the chest, tumor elimination, and the cardiovascular system. He advanced urology; the surgery of the vegetative nervous and endocrinological systems; and anesthesiology. He built up a school of his followers and engaged in pedagogical and organizational activities. Today, there is a Gertsen Moscow Research Institute of Oncology (3 Second Botkinsky Passage).

Petr Kashchenko (Chapter 5).

Iosif A. Kassirsky (1898–1971) [7/15] was a physician and hematologist whose medicinal education was interrupted by the civil war. He was an innovator physician in broad areas of therapeutics. He made significant contributions to saving lives in WWII, especially through his activities in hematology. His bust over his grave is by V. V. Miklashevskaya.

Vasily K. Khoroshko (1881–1949) [2/14] worked in neuropathology and did research of the vegetative nervous system, epilepsy, and brain damage. He introduced methodological innovations in medicine.

Nikolai V. Konovalov (1900–1966) [6/26] was a neuropathologist whose research focused on progressive degenerative and inherited pathologies of the nervous system.

Andrei V. Lebedinsky [6/14] (Chapter 4).

Aleksandr V. Leontovich (1869–1943) [2/41] was a physiologist and histologist. His principal interest was in the degeneration and regeneration of the nervous tissue.

Sergei R. Mardashev (1906–1974) [4/56] conducted biomedical research of insulin and in enzymology. For a long time he was in charge of biochemistry at the First Moscow Medical Institute.

Vasily I. Molchanov (1868–1959) [5/34] was a professor of pediatrics at the First Moscow Medical Institute. He co-authored a monograph about pediatrics.

Aleksandr Myasnikov [6/29] (Chapter 5).

Vasily V. Parin [7/16] (Chapters 4 and 5).

Leonid Persianinov [9/3] and Boris Petrovsky [10/9] (Chapter 5).

Leonid S. Persianinov (1908–1978) is depicted on his gravestone with a newborn baby in his hands. From 1958, he was in charge, consecutively, of three departments and a research center of obstetrics and gynecology. His interests included diverse problems of birth and the trauma that may happen during birth.

Petr P. Popov (1888–1964) [3/57] was an epidemiologist at the medical institute of the Azerbaijani Academy of Sciences. The relief on his gravestone is by A. Vrubel.

Boris S. Preobrazhensky (1892–1970) [7/15] was an otolaryngologist and held professorial appointments in medical schools in Moscow. He introduced innovations for the treatment of his patients.

Nikolai Priorov [8/11] and Yury Senkevich [10/8].

Nikolai N. Priorov (1885–1961) was a specialist in traumatology and orthopedics. For a short time he was deputy minister of health and he served as a physician of the Soviet Olympic team in 1952 in Helsinki.

Ivan Sechenov [4/38] (Chapter 5).

Nikolai Semashko [1/46] (Chapter 5).

Yury A. Senkevich (1937–2003) started as a physician, but turned into a journalist who anchored popular TV shows. His background in science enhanced the depth and value of his science reporting and popularization of science. He gained experience in military medicine and in the biomedical aspects of space exploration. He participated in expeditions and studied the human behavior under extreme conditions. The motifs of his tombstone indicate his interests in natural history.

Evgeny K. Sepp (1878–1957) [5/6] researched the nervous system and was head of the department of the pathology of the nervous system at the First Moscow Medical Institute.

Petr G. Sergiev (1893–1973) [4/36] graduated as a physician from the medical school of Kazan University. He fought in the Civil War as a Red Army volunteer. He conducted party work and for some time he was deputy minister of health. His main contribution was as an epidemiologist and during WWII he directed the Soviet anti-malaria measures.

Konstantin Shkhvatsabaya [8/2] and Valery Shumakov [10/10] (Chapter 5).

Konstantin Ya. Shkhvatsabaya (1896–1960) was a specialist of internal medicine. His sons, the cardiologist Igor K. (1928–1988), and the sport cardiologist and specialist of internal medicine, Yury K. (1927–1994), as well as Konstantin's epidemiologist wife, Tatiyana V. Shkhvatsabaya (1905–1990), are buried in the same grave.

Mikhail A. Skvortsov (1876–1963) [8/25] was an anatomist and the founder of pediatric pathological anatomy in Russia. Between 1911 and 1953, he worked as pathologist at the Moscow Clinical Pediatric Hospital. He was also professor of pediatrics at Moscow medical schools. He created a unique collection of preparations related to a broad variety of pathological conditions in pediatrics. In the same grave, there are the remains of his wife, the renowned physician Ekaterina A. Kost (1888–1975). She graduated in 1913 from the Higher Courses for Women and rose to the rank of professor. Her specialties were hematology and laboratory analyses. Her monographs covered a wide range of topics in medicine.

Efim I. Smirnov [11/1] (Chapter 5).

Iosif L. Tager (1900–1976) [9/2] was a specialist of the medical uses of X-rays. He was an active participant at the front during WWII. Before and after the war he worked in clinics and research institutions. He was among those mostly Jewish doctors and professors who were accused of conspiracy to assassinate Soviet leaders. This was the so-called "doctors' plot" not long before Stalin's death. Tager was fired from his job. When this turmoil was over, he resumed his research and clinical work. His main area of activities was diagnostics.

Evgeny M. Tareev (1895–1986) [6/24] (Chapter 5). Galina A. Raevskaya (1900–1966), buried together with him, was his second wife (his first wife died in 1920). Raevskaya was a renowned cardiologist and had the rank of professor. One of their two daughters, Irina E. Tareeva (1931–2001), buried also in the same grave, was a physician, specializing in nephrology and internal medicine. She was the head of a department at the Sechenov Medical University.

Vladimir D. Timakov [9/2] (Chapter 5) and Vladimir N. Vinogradov [3/49].

Vladimir N. Vinogradov (1882–1964), a specialist of internal medicine and cardiologist, worked at the First Moscow Medical Institute during his entire career. Upon his death, the clinic of internal medicine was named after him. Vinogradov introduced new approaches in the treatment of cardiovascular diseases and keenly followed the international developments in his field. From 1943, he was chief of internal medicine of the Kremlin hospital. In November 1952, he was arrested in connection with the so-called "doctors' plot." He was accused of the killing of Andrei A. Zhdanov, one of Stalin's top lieutenants, of American espionage, and of other crimes. Stalin died on March 5, 1953, and Vinogradov was freed on April 4. Vinogradov's two sons were buried in the same grave: Georgy (1910–1988), a polymer chemist, and Vladimir (1920–1986), a surgeon. The gravestone was created by M. K. Anikushin.

Aleksandr A. Vishnevsky [9/1] (Chapter 5). His father, Aleksandr V. Vishnevsky (Chapter 5) is buried elsewhere in this cemetery.

Sergei S. Yudin [2/32] (Chapter 5).

Boris I. Zbarsky (1885–1954) was a biochemist who studied in Geneva after he had been prevented from completing his education in tsarist Russia. He returned to Russia in 1912 and served in the Russian army in WWI. In the 1920s, Zbarsky assisted establishing research institutes. When Lenin died, Zbarsky assisted the anatomist Vladimir P. Vorobiev in the conservation of Lenin's body. After Vorobiev's death in 1937, Zbarsky was appointed to be in charge of the preservation of Lenin's body and of the laboratory serving the Lenin Mausoleum. He solved the problems of the transportation of Lenin's body during the war years first to move it out of Moscow and subsequently to return it. His merits were recognized with the highest awards. He was active in persecuting Roskin and Klyucheva (Chapter 5). However, in 1952, he also became a victim of terror. He was arrested and accused of underplaying Stalin's role in the revolution in his writings about Lenin. His case was under the umbrella persecution of the so-called "doctors' plot." The charge against him was that he was a German spy, but his case was never tried. He was kept in jail for months even after Stalin's death. After he was freed he spent his last years as professor at the First Moscow Medical Institute.

Dmitry Zhdanov [7/16].

Dmitry A. Zhdanov (1908–1971) [7/16] was an anatomist and morphologist and head of the department of anatomy of what is now the Sechenov Medical University, from 1956 until his death.

Earth Scientists and Explorers

Nikolai N. Baransky (1881–1963) [8/39] was one of the founders of geographical economics in the Soviet Union.

Valery L. Barsukov (1928–1992) [10/7] was a geochemist. He examined rock samples from planet Venus. His research interest was in the geochemical processes of rock formation.

Dmitry Belyankin [4/25] and Evgeny Fedorov [9/8].

Dmitry S. Belyankin (1876–1953) pioneered technical petrography and led expeditions to the Ural Mountains, the Caucasian Mountains, and to the northern regions of the country.

Aleksei A. Borisyak [3] (Chapter 3).

Aleksandr Borzov (Chapter 3).

Evgeny K. Fedorov (1910–1981) geophysicist was the director of what is today the hydro-meteorological service including monitoring the environment. In 1937–1938, he participated in the four-member North Pole expedition whose leader was Ivan D. Papanin. The other two members were Ernst Krenkel and Petr Shirshov (see below).

Aleksandr Fersman [3/44] (Chapter 3).

Ivan Gubkin [1/46] (Chapter 4).

Aleksandr S. Ilichev (1898–1952) [1/2] was a geologist who greatly contributed to the successes of petrography in the Soviet Union.

Grigory N. Kamensky (1892–1959) [5/40] was a hydro-geologist whose principal research was in theoretical studies.

Left: Ilya Kibel and Ekaterina Blinova [7/12]. *Right*: Nikolai Kochin and Pelageya Kochina [4/40].

Ilya A. Kibel (1904–1970), a mathematician by training, specialized in hydro-mechanics and meteorology and his mathematical approach reformed weather forecasting. His methods received enormous attention when they were successfully applied during the blockade of Leningrad. The lifeline for the city was over Lake Ladoga and it was vital to know what to expect about the freezing of the lake. Kibel's methods proved a lifesaver. The geophysicist Ekaterina N. Blinova (1906–1981), Kibel's wife, suggested a mathematical method for weather forecast by integrating the equations describing the conditions of the atmosphere. Her research, just as Kibel's, reached back to the teachings of the legendary Russian astrophysicist and mathematician Aleksandr A. Fridman (Friedmann, 1888–1925).

Nikolai E. Kochin (1901–1944) and Pelageya Ya. Kochina (1899–1999) were both academicians. Kochin, graduated as a mathematician and physicist, was one of the founders of dynamical meteorology. He had also results in gaseous dynamics and the theory of shock waves in compressible liquids. Kochina (née Polubarinova) studied first at the Higher Courses for Women, later at Petrograd University. Aleksandr Fridman was among her professors. Eventually, Kochina became a renowned specialist in hydrodynamics. From 1970, she was head of the section of mathematical methods in mechanics of the Institute of Problems of Mechanics.

Ernst T. Krenkel (1903–1971) [7/right side18], radio operator, was participant in the North Pole expedition led by Papanin.

Petr Kropotkin (Chapter 3).

Aleksandr Mazarovich (Chapter 3).

Evgeny Milanovsky (Chapter 3).

Stepan I. Mironov (1883–1959) [5/33] was a geologist and oil specialist. He was a strong supporter of the theory of the organic origin of oil. He was a professor at what is now Gubkin University.

Vladimir Obruchev [1/30] (Chapter 3). The sculptor of the memorial on Obruchev's grave was Z. M. Vilensky.

Ivan D. Papanin (1894–1986) was an explorer and, eventually, a rear admiral of the Navy. He has already figured above as the leader of the North Pole expedition in 1937–1938 and he was in charge of the first permanent polar station "North Pole." In recognition of the success of the expedition, Papanin and Krenkel received the scientific degree of doctor of geographical sciences (DSc). From 1951, Papanin was in charge of the section of maritime expeditions of the Academy of Sciences.

Mariya V. Pavlova [1/33] (Chapter 8).

Aleksandr V. Peive (1909–1985) [9/2] was a geologist. He researched tectonics and magnetism, especially of the oceanic earth crust.

Vladimir Rzhevsky [10/7] and Otto Schmidt [1/42].

Vladimir V. Rzhevsky (1919–1992) lost his parents early in his childhood. He graduated from the Moscow Mining Institute in 1941. After his participation in WWII he continued his career at the same institute becoming its rector in 1962. He modernized education at this institute and contributed to the theory and practice of applications of petrography.

Aleksandr A. Saukov (1902–1964) [6/11] was a biochemist involved in geology and mineralogy. He participated in geological expeditions and contributed to the utilization of discoveries in petrography.

Otto Yu. Schmidt (1891–1956) graduated from Kiev University and became an explorer with a good background in mathematics, geophysics, geography, and astronomy. In 1928 he participated in a pioneering expedition to the Pamir Mountains. In the early 1930s, Schmidt led expeditions to the North, including the one in 1932 when the expedition sailed from Arkhangelsk to the Bering Strait. This was a milestone for the development of the far eastern regions of the Soviet Union. He became director of the Arctic Institute and later, director of the Authority of the Northern Sea Route. He served as vice president of the Academy of Sciences between 1939 and 1942. His tombstone is the work of S. T. Konenkov.

Dmitry I. Shcherbakov (1893–1966) [6/31], geologist, mineralogist, and geochemist, studied in St. Petersburg/Petrograd/Leningrad and began his career at the Authority of Mineral Resources. Another venue for him was the Radium Institute and he moved up on the academic ladder. He participated in numerous expeditions with great success in finding new resources and learning about far-away places in the country. Eventually, a distinguished goal of the expeditions was locating occurrences of uranium ore. There was great success in this as well. When a new Institute of Geochemistry was established in Leningrad, Shcherbakov joined it and moved to Moscow with the reorganized Institute of Geosciences in the mid-1930s. His contributions to the uranium project were much appreciated by the state and by the academy leadership.

Lev. D. Shevyakov (1889–1963) [8/25] was a professor of mining sciences, and after other locations he worked at the Moscow Mining Institute for the last three decades of his life.

Petr P. Shirshov [4/23] and Aleksandr Sidorenko [9/9].

Petr P. Shirshov (1905–1953) was a member of the Papanin expedition and other expeditions, including the one that charted various characteristics of Novaya Zemlya (New Land), an archipelago in the Arctic Ocean. In the 1930s he worked at the Arctic Institute. When the director of the institute, Rudolf L. Samoilovich was arrested, Shirshov took over the directorship. Samoilovich was executed in 1939 and rehabilitated in 1957. Between 1942 and 1948, Shirshov was minister of the Maritime Ministry and from 1946, the first director of the Institute of Oceanography (36 Nakhimovsky Avenue). As a high-ranking maritime official Shirshov was involved in the transportation of prisoners to inhuman labor camps.

Aleksandr V. Sidorenko (1917–1982) [9/9] graduated as a geologist in 1940 from Voronezh University. Following a few years of doing scientific research, his political career took off and he was minister of geology of the Soviet Union between 1962 and 1975 and then vice president of the Academy of Sciences.

Nikolai N. Slavyanov (1878–1958) [5/27] hydro-geologist graduated from the St. Petersburg Mining Institute in 1908. He had various functions related to geology and taught at institutions of higher education in Leningrad and Moscow. Between 1947 and 1956, he was director of the laboratory of hydro-geology of the Academy of Sciences. His main research interest was in mineral waters in the most diverse regions of the Soviet Union.

Aleksandr Terpigorev [5/30] and Vladimir Vernadsky [3/45] (Chapter 3).

Aleksandr M. Terpigorev (1873–1959) graduated from the St. Petersburg Mining Institute and in Soviet times he became one of the most authoritative mining specialists in the country. From the early 1920s, he was a professor and department head at the Moscow Mining Institute. From 1938 to 1948 he was the head of the department of mining machines and from 1950 until his death, of the treatment of layered occurrences of mineral resources. Parallel to his positions at the Institute, Terpigorev occupied various leading state positions related to mining.

Aleksandr P. Vinogradov (1895–1975) [9/1] studied chemistry at Leningrad University. He was a geochemist and a close associate of Vladimir Vernadsky, founded the first department of geochemistry in the country at Moscow University, and directed the Vernadsky Institute of Geochemistry and Analytical Chemistry in Moscow (19 Kosygin Street). He was included in the nuclear program from the late 1940s and directed the development of highly sensitive techniques of analytical chemistry to ensure the production of pure materials when they were needed. Following the test explosion of the first Soviet nuclear device, Vinogradov was among those who received the highest governmental recognitions.

Lev A. Zenkevich (1889–1970) [7/11] was a biologist with a special interest in the biology of seas and oceans. He graduated from Moscow University and continued there, eventually, as a professor. He participated in maritime expeditions. Throughout his career he was active in institutions dealing with marine biology. His entire professional career was related to Moscow University.

Nikolai Zubov (Chapter 3).

Statue of Andrei Sakharov (G. V. Pototsky, 2008) at the Muzeon Sculpture Park.

8

Keep Walking

Moscow is an inexhaustible treasury of memorials on science and scientists. We have compiled a variety of further examples to illustrate its wealth and diversity. A visitor in Moscow may find many more relevant examples.

Andrei Sakharov (Chapter 1) was not only one of the foremost physicists of his time; he was also a human rights activist. The Soviet authorities had a difficult time deciding how to silence the "father of the Soviet hydrogen bomb" and one of the most decorated citizens of the Soviet Union. Finally, they exiled him to Gorky (before and now, Nizhny Novgorod) in January 1980. He was kept there under constant harassment until the end of December 1986.

Statues of Felix Dzerzhinsky (E. Vuchetich, late 1950s) and Andrei Sakharov (G. V. Pototsky, 2008) in the Muzeon Sculpture Park.

There are two sculptures in close proximity in the Muzeon Sculpture Park (see below) with a peculiar relevance to each other. Felix E. Dzerzhinsky (1877–1926) was a Soviet leader, best known as the founder of the infamous Soviet security police, which had a variety of names over the years, among them, Cheka, GPU, NKVD, and KGB. Dzerzhinsky's statute used to stand in the center of Lubyanka Square, in front of the headquarters of the secret police, radiating force, power, and authority. The Sakharov sculpture on its own could be representing an elderly man sunbathing. However, Sakharov with Dzerzhinsky in the background reminds us of the heroic struggle the defiant Sakharov had taken up with inhuman repression. This struggle must have seemed hopeless at the time yet this fragile man defeated the seemingly invincible powers that be.

The name Lubyanka has become synonymous with the secret police. They kept many innocent people, among them scientists, incarcerated there. Today, at the edge of Lubyanka Square, there is a modest memorial to the victims of the repressive Soviet system.

A stone from the GULAG in the lower right corner of the image. It is in remembrance of the victims with the infamous Lubyanka in the background — the headquarters of the secret police. GULAG is the acronym of Glavnoe Upravlenie Lagerei — Main Authority of Camps.

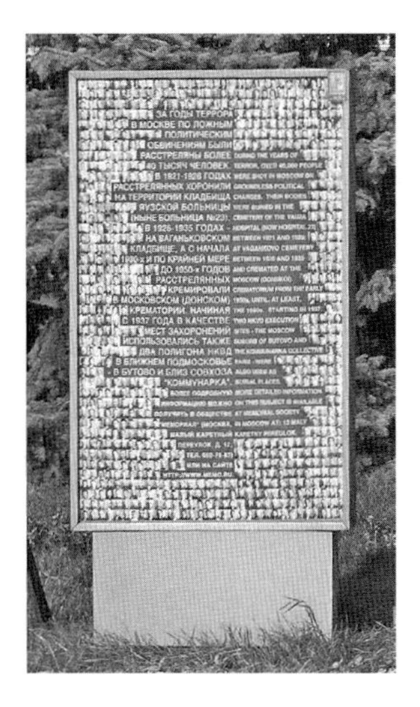

Left: Partial list of the crimes embedded among portraits of victims. *Right*: Prison photo of Lev Landau, who spent one year at the Lubyanka (courtesy of the late Boris Gorobets).

Andrei Sakharov

> "… my fate was in some sense extraordinary …
> I note — and not for false modesty, but for I want to be exact —
> that my fate turned out to be grander than my personality.
> I was merely striving to stay on the level of my own fate …"
>
> *Andrei Sakharov, 1988*

The quoted statement is exhibited at the Sakharov Archive at 48B Zemlyanoi Val Street, Sakharovs' last residence. A memorial plaque marks the façade, noting, "Sakharov lived here."

Dropped a few meters back from the busy thoroughfare, at 57 Zemlyanoi Val Street, there is a Sakharov Center with a detached exhibition hall. The small public garden — the Sakharov Garden, between Zemlyanoi Val Street proper and the Sakharov Center — conveys calm, but remembers the past with a piece of the Berlin Wall.

Left: A piece of the Berlin Wall in the Sakharov Garden. *Right*: A portion of the Wall in Berlin (2016) with Sakharov's portrait and an inscription in German, "Thank you Andrei Sakharov."

The Wall was erected in 1961 to prevent the citizens of East Germany from defecting to the West. It was dismantled in 1989. The division of Germany came about as a consequence of WWII and lasted until the political changes in 1989. The Berlin Wall had become a symbol of the inhuman nature of the Soviet-style system.

Sakharov Center and its detached exhibition hall at 57 Zemlyanoi Val Street.

Two Sakharov busts by Peter Shapiro; at the Sakharov Center in Moscow (*left*) and in front of the Russian House at 1800 Connecticut Avenue, Washington, DC (*right*).

The Sakharov Archive and the Sakharov Center are full of memorabilia. The Archive has display windows that inform the visitors, often groups of schoolchildren, in cozy surroundings. The Center has larger halls and display windows and is geared to engaging larger groups of visitors.

Museum of Modern Art

The Moscow Museum of Modern Art at 25 Petrovka Street is one of the five venues of this Museum. Modern art used to be persecuted in Soviet times, but still, private collections existed. This Museum was formed in 1999 based on the private collection of the renowned Georgian-born Moscow artist Zurab K. Tsereteli (1934–). Some of its artifacts represent scientists.

Vladimir Minaev, "Interrogation of Nikolai I. Vavilov," 1987.

Yury Orekhov, "Andrei D. Sakharov," 1990s.

Zurab Tsereteli, "In memory of the victims of oppression, N. Vavilov, M. Tsvetaeva, B. Pilnyak, and D. Shostakovich," 1994. This is a composition of 5 parts. *Left*: the first four parts; *right*: the fifth part of the composition; the door of the prison cell from the inside of the cell with shoes and hats with earflaps that the prisoners left behind as they were led to their execution.

Tsereteli dedicated a five-part memorial to the victims of oppression. In particular to Nikolai I. Vavilov (Chapters 1 and 6); Marina I. Tsvetaeva (1892–1941), the poet who committed suicide; Boris A. Pilnyak (1894–1938), the writer who was arrested on false charges, sentenced to death, and was executed; and Dmitry D. Shostakovich (1906–1975), the composer and pianist whose work was denounced by Soviet officialdom in 1936 and in 1948.

There is also a Tsereteli Art Gallery at 19 Prechistenka Street, which opened in 2001 in the former mansion of the Dolgorukov family. It is both a museum and an exhibition hall.

Statue of Zhores Alferov (Z. K. Tsereteli) in the Tsereteli Art Gallery (courtesy of Olga Dorofeeva).

Zhores I. Alferov (1930–2019), a native of Belarus, studied at the Minsk Polytechnic Institute and graduated from the Leningrad Electrotechnical Institute in 1952. From 1953, he worked at the Ioffe Institute of Physical Technology, and rose through the ranks to the directorship (1987–2003). He served also at various leadership positions of academic and educational institutions. His main field of research was electronics and he was co-recipient of the Nobel Prize in Physics in 2000 for his discoveries in semiconductor hetero-structures and their applications. He was one of the ten academicians who in 2007 signed a letter of protest to the President of Russia against the clericalism in Russian society.

The Muzeon Park of Arts

The Muzeon is a whole complex of sculpture gardens, exhibition halls, and other institutions. Its official address is 2 Krymsky Embankment, and the best approach to it is walking from the subway station Oktyabrskaya. The original idea of the sculpture garden in 1992 was to collect the monuments of the Soviet era that no longer should be exhibited in public space, but should not be destroyed either. Their preservation could help maintain societal memory and they may also represent artistic value. However, the contents of the sculpture garden have changed gradually. They have not stayed limited to the memorials of a discredited historical era, while not all the memorials of discredited politicians have been removed from their original places.

M. F. Baburin, "Triumph of Labor: Science and Art," a Soviet-style sculpture at the main entrance to the Muzeon.

From left to right: Busts of the Persian polymath, Avicenna; Aleksandr Fersman (E. B. Preobrazhenskaya, 1991); and Albert Einstein (Z. M. Vilensky, 1979).

Two statues of M. V. Lomonosov (L. M. Baranov, 1943, *left*, in the Tretyakov Filial, see below, and 1980, *right*, in the open-air sculpture park).

A section of the Muzeon exhibits monuments of principal ideologues of Marxism and political leaders of the Soviet Union. There is a monumental sculpture of Stalin, with his face disfigured by anti-Stalin protesters. The positioning of this sculpture is especially powerful: it stands before the memorial of his victims.

S. D. Merkulov's "Stalin, 1938" in front of E. I. Chubarov's "Victims of Repression, 1998."

Albert Einstein and Niels Bohr by Vladimir Lemport (photograph by Beau Bernatchez; courtesy of Igor and Katya Gomberg).

One of the most intriguing sculptures of scientists in the Muzeon Park used to be the statue of Albert Einstein and Niels Bohr sitting together on a bench; both of them smoking their pipes. In 2014, this sculpture was purchased and moved to the sculpture garden of the Turn Park Art Space at 2 Moscow Road in West Stockbridge, Massachusetts.

Peter the Great

Monument of Peter the Great (Z. Tsereteli, 1997), at the point where the River Moscow and the Vodo-otvodny Canal meet, as viewed from the Muzeon.

From the Muzeon Park, there is a clear view of the Peter the Great (1672–1725) memorial. He ruled between 1682 and 1725. Peter attempted building a scientifically and technologically based empire. He was more determined to keep pace with progress than his predecessors and his successors prior to the Soviet rule. He founded the Russian Academy of Sciences in 1724.

Tretyakov Filial

There is the Central House of the Artist in the Muzeon and more than half of its space belongs to the Annex of the State Tretyakov Gallery — the Tretyakov Filial — dedicated to 20th century art. Close to its entrance stands a copy of the sculpture of a man turning a sword into a plowshare symbolizing the transition of nuclear science from weapons to peaceful energy production. The original was a gift of the Soviet Union to the United Nations.

E. Vuchetich's statue in the North Garden of the UN Headquarters in New York (*left*) and its copy in the Muzeon Park of Arts (*right*).

From the Tretyakov Filial we show here three more pieces of art.

Left: Vera T. Mukhina, "Academician A. N. Krylov," 1945. *Right*: Nikolai B. Nikogosyan, "Petr L. Kapitsa," 1956. For Krylov and Kapitsa, see Chapter 1.

Maxim K. Kantor, "Chernobyl, 1987."

Chernobyl, just like Lubyanka, is more than the name of a location. It signifies the world's largest nuclear catastrophe and refers to something even more sinister than the catastrophe itself. It happened already when Mikhail Gorbachev's policy of openness was ostensibly in effect, but the response of the Soviet organs reverted back to the old Soviet style reactions. They delayed informing the public and initially belittled what happened.

Konenkov Studio-Museum

Sergei T. Konenkov (1874–1971) created portraits of scientists, among others. He spent two decades of his life in the United States and was ordered back to the Soviet Union in 1945. We showed his busts of biomedical scientists in our previous book, *New York Scientific*, including Simon Flexner, Phoebus A. Levene, Hideyo Noguchi, and William H. Welch, all of The Rockefeller University for Medical Research.

From 1947, Konenkov lived at the corner of 17 Tverskaya Street and 28 Tverskoi Boulevard, which is now the Konenkov Studio-Museum (with the entrance from the boulevard side).

Left: Relief on the façade of the Konenkov Studio-Museum. *Right*: Konenkov's grave in the Novodeviche Cemetery.

Left: Old Tatar reading, 1895, symbolizing literacy and the thirst for knowledge. *Right*: Ivan P. Pavlov, 1930, the 1904 Nobel laureate in Physiology or Medicine.

Albert Einstein as a bust and as a miniature statue by Konenkov, both in 1935. The artist's portrait is seen behind the Einstein bust.

Vadim Sidur Museum

Vadim A. Sidur (1924–1986) was an avant-garde sculptor whose sculptures could not be exhibited from 1950 during the rest of his life. In the 1970s though at least two of his creations appeared in public due to the affinity some scientists felt towards his art. Both these two pieces were tombstones at the Novodeviche Cemetery (over the graves of academicians Frumkin and Tamm, Chapter 7).[1]

Vadim Sidur's portrait of Vitaly L. Ginzburg.

The Moscow State Vadim Sidur Museum at 37 Novogireevskaya Street from the outside and its main exhibition hall.

[1] In 2015, several of Sidur's works were smashed at an exhibition in Moscow by protesters who considered his art to be blasphemous.

Nikogosyan Art Gallery

In addition to Petr Kapitsa's bust (see, above), Nikolai B. Nikogosyan (1918–2018) has sculpted the busts of other academicians and two other Nobel laureate physicists. They are on display in his art gallery at 19 Bolshoi Tishinsky Street.

Busts of John D. Cockcroft (*left*) and Willis E. Lamb (*right*) by Nikolai Nikogosyan; courtesy of David Nikogosyan.

John D. Cockcroft (1897–1967) was a British physicist who shared the Nobel Prize in 1951 with Ernest Walton for transmuting atomic nuclei by means of bombarding them with accelerated fundamental particles.

Willis E. Lamb (1913–2008) was an American physicist who was awarded the Nobel Prize in 1955 for uncovering the fine structure of the hydrogen spectrum.

Busts by Nikolai Nikogosyan (*from left to right*): Viktor A. Ambartsumyan, Samvel S. Grigoryan, and Aleksandr Yu. Ishlinsky (courtesy of Olga Dorofeeva).

Viktor A. Ambartsumyan (1908–1996) was a Soviet-Armenian astrophysicist. He graduated from Leningrad University and did post-graduate studies at the Pulkovo Observatory. He was a long-time president of the Armenian Academy of Sciences. He was among the founders of theoretical astrophysics and developed theories about the birth and evolution of stars and galaxies. He initiated the astrophysical observatory in Byurakan, Armenia. He was awarded the Gold Medal of the Royal Astronomical Society (UK) and the Bruce Medal of the Astronomical Society of the Pacific.

Samvel S. Grigoryan (1930–2015) was a Soviet-Armenian mathematician. He graduated from Moscow University and continued his career there, eventually as director of the Institute of Mechanics. His research concerned the properties of condensed-phase materials.

Aleksandr Yu. Ishlinsky (1913–2003) was an applied mathematician. He was a graduate and then a professor of Moscow University. In addition, he had a number of other jobs, involving teaching, research, and industrial applications. He was the first director of the Institute of Mechanics of Moscow University. He was a member of many international learned societies.

Darwin Museum

The State Darwin Museum at 57 Vavilov Street.

A number of Moscow museums have already been mentioned above and in previous chapters, such as the Kapitsa Museum and the Nikolai Vavilov

Museum (Chapter 1); the Earth Science Museum (Chapter 3); the Polytechnical Museum and the Cosmos Museum (Chapter 4); the Museum of the History of Medicine and the Museum of Psychiatry (Chapter 5); and the museums at the Timiryazev Academy (Chapter 6). The Moscow State Darwin Museum is another natural history museum. Aleksandr F. Kots (Alexander E. Kohts, 1880–1964), a German biologist, founded it. He immigrated to Russia in 1907 to teach at the Moscow Higher Courses for Women and later also at Moscow University.

Expositions at the Darwin Museum.

Statues of Charles Darwin (V. A. Vatagin) and Jean-Baptist Lamarck with his daughters (courtesy of Olga Dorofeeva) at the Darwin Museum.

Other Museums

Timiryazev State Biological Museum and its Ivan Pavlov statue (D. Ryabichev). Courtesy of Olga Dorofeeva.

The **Timiryazev Biological Museum**, 15 Malaya Gruzinskaya Street, was founded in 1922 and has been at its present magnificent location since 1934. It is the second most popular museum among families with children (the first is the Darwin Museum).

The Fersman Mineralogical Museum at 18 Leninsky Avenue (under renovation in September 2016) and one of its exhibition halls (courtesy of Olga Dorofeeva).

The **Fersman Mineralogical Museum** was founded in 1716 in St. Petersburg and moved to Moscow in 1934. It has some 135-thousand specimens. It was reopened recently, freshly renovated.

The Vernadsky Geological Museum at 11 Mokhovaya Street and its Vernadsky bust (Z. M. Vilensky, courtesy of Olga Dorofeeva).

The **Vernadsky Geological Museum** was founded in 1755 and is the oldest Moscow museum. More than a venue for exhibiting its thousands of geological specimens, it is also an institution for the popularization and dissemination of science.

Two scenes in the Orlov Museum of Paleontology (courtesy of Olga Dorofeeva).

The **Yu. A. Orlov Museum of Paleontology** at 123 Profsoyuznaya Street is part of the Borisyak Institute of Paleontology. It presents the evolution of the living world on our Planet. The Museum traces back its origin to 1714; its current venue opened in 1987.

Top row from left to right: Viktor Amalitsky, Nikolai Andrusov, Georges Cuvier, and Charles Darwin (Chapter 3). *Bottom row*: Louis Dollo, Afrikan Krishtofovich, Melchior Neumayr, and Mariya Pavlova. Courtesy of Olga Dorofeeva.

There are a number of reliefs on display honoring international paleontologists, such as Georges Cuvier, Charles Darwin, and Louis Dollo, and Russian ones, among them, Aleksei A. Borisyak, the founder of the Institute of Paleontology and Yury A. Orlov, one of the founders of this Museum.

Viktor P. Amalitsky (1860–1917) studied in St. Petersburg and worked at Warsaw University. He conducted explorations in Northern European Russia and discovered near-perfect fossil remains of amphibians and reptiles.

Nikolai I. Andrusov (1861–1924) graduated from Novorossiisky University. Beside Russian universities, he taught at the Sorbonne and at Charles University in Prague. He collected fossil specimens and explored deep-sea conditions.

Georges Cuvier (1769–1832), "the father of paleontology," investigated the changes in anatomy over millennia through his studies of fossils.

Louis Dollo (1857–1931) was a Belgian scientist, a pioneer of paleontology and one of founders of paleo-biology.

Afrikan N. Krishtofovich (1885–1953) founded botanical paleontology, the science of fossil plants, in Russia. His career was interrupted when he was arrested in 1930, as part of the attack on the Academy of Sciences. After a year and a half in prison he was sentenced to 5 years of exile, but allowed to lecture in Sverdlovsk. Upon his return to Moscow he continued his research, which earned both Soviet and international recognition.

Melchior Neumayr (1845–1890) was a German paleontologist who studied in Munich and worked mostly in Austria. His two-volume *Erdgeschichte* (History of Earth) appeared in Russian translation.

Yury A. Orlov (1893–1966) was a zoologist and paleontologist. He graduated from Petrograd University. He taught at a number of institutions of higher education before he was appointed head of the department of paleontology of Moscow University in 1943. Working in paleontology was his childhood dream. From 1945 to the end of his life, he directed what is today the Borisyak Institute of Paleontology. He was a vocal supporter of Lysenko, but in 1955 he changed sides and joined Lysenko's critics.

Mariya V. Pavlova (1854–1938) graduated from the Sorbonne. She worked in the geological museum of Moscow University and from 1911 she had a professorial appointment at Sanyavsky University. From 1919, she was at Moscow University, the first female professor of the institution. Her multifaceted research was in natural history. She was married to the renowned paleontologist Aleksei Pavlov (Chapter 3).

Petr P. Sushkin (1868–1928) studied at Moscow University under Mikhail Menzbir's (Chapter 2) mentorship. Sushkin was a zoologist and paleontologist and Yury Orlov was one of his disciples.

Nikolai N. Yakovlev (1870–1966) was one of the founders of paleontology in Russia and president of the Paleontological Society between 1916 and 1940.

Other Institutions

The principal building of the Tsitsin Main Botanical Garden and Nikolai V. Tsitsin's memorial plaque in its front (courtesy of Olga Dorofeeva).

The **N. V. Tsitsin Main Botanical Garden** in Moscow was founded in 1945, shortly before the end of WWII. This symbolized the victory of the Soviet Union in the war. It is located adjacent to the All-Russia Exhibition Center (VDNKh) and the total area of the Garden is over 331 hectares with 18-thousand different plants. In 1991, the Garden was named after the botanist Nikolai V. Tsitsin (1898–1980). He held high administrative positions related to the agricultural and biological sciences.

The Moscow Zoo Park at Bolshaya Gruzinskaya Street (courtesy of Olga Dorofeeva).

The **Moscow Zoo Park** was opened in 1864 initiated by three professors of Moscow University, Karl F. Rulie, Anatoly P. Bogdanov, and Sergei A. Usov. One of its later directors was the biologist Aleksandr Kots who had founded the Darwin Museum. The Zoo Park survived difficult times, the revolutions, the Civil War, and Stalin's Terror. Even during the five years of WWII, it stayed in operation and had millions of visitors.

The statue of a bee on the backdrop of the Eco-center and a memorial plaque of Aleksandr V. Serzhantov (courtesy of Olga Dorofeeva).

One of Moscow's numerous parks, the **Kuzminsky Park**, at 10 Kuzminskaya Street, is called also the Kuzminky-Lyublino Park of Natural History. It has an "Eco-center" established in 2003 for the dissemination of ecological information. There is a bronze statue of a bee in the Park, a symbol of diligence, well-being, and success. This Eco-center developed from a museum-educational center for bee-keeping. The ecologist Aleksandr V. Serzhantov was instrumental in developing the Eco-center.

Moscow Planetarium

The Moscow Planetarium, 5 Sadovaya-Kudrinskaya (courtesy of Olga Dorofeeva).

The Moscow City Council decided to establish a planetarium in 1927. The foundation stone was laid in 1928 and the planetarium was opened to the public in 1929. It is now a venue for the dissemination of knowledge, a museum, and a research institute as well.

Mobius Bands

We present here two artistic appearances of the Mobius band in Moscow. The Mobius band is a two-dimensional surface with having only one side. It exists in physical reality although it is somewhat against common sense. We can form a Mobius band by twisting a paper strip by 180 degrees about its axis and attaching the two ends.

Left: A monumental stylized Mobius band (L. Pavlov, L. Zharenova, and V. Vasiltsov, 1980) on the façade of the Central Institute of Economics-Mathematics, 47 Nakhimovsky Prospekt. *Right*: Black Mobius (A. Nalich) at 21 Komsomolsky Prospekt. Some viewers may recognize a nude figure when looking at the sculpture from a certain angle.

Metro Stations

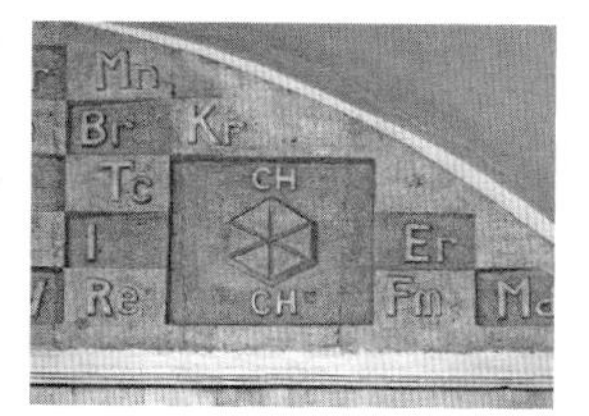

Murals at the Mendeleevskaya Metro station.

Mendeleevskaya station. At the end of its hall there is a mural with parts of the periodic table of the elements and a Mendeleev portrait. It is intriguing that some cartoons in this mural look like benzene structures as if representing features of the resonance theory. This theory was anathema to official Soviet ideology around 1950 (Chapter 2).

The Mendeleevskaya station was designed by Nina A. Aleshin and Natalya K. Samoilova. They had consulted associates of Mendeleev University (Chapter 4) before they finalized their design as they wanted it relate to Dmitry Mendeleev. The station opened in 1988. The lighting arrangement reminds the scientist visitors of crystal structures and the murals remind them of electron density distributions in atoms and molecules.

Detail of the mural of Tmiryazev lecturing at the Timiryazevskaya Metro station.

Timiryazevskaya station has a mural of marble of the biologist Timiryazev talking to young people of the agricultural academy, the Petrovsky Academy in his time. The artists M. A. Shorchev and L. K. Shorchev created the mural in 1991.

Sculptures at the Ploshchad Revolutsii Metro station: Children playing with a globe, with an airplane model, and a girl reading.

Ploshchad Revolutsii (Revolution Square) station presents a total of 76 sculptures depicting 20 different scenes of which we chose three for presentation. These three display young people learning, playing, and reading. A collective of sculptors headed by Matvei G. Manizer created the sculptures in the years 1936–1938.

Printer Ivan Fedorov

Statue of the Printer Ivan Fedorov by Sergei M. Volnukhin, 1909.

Ivan Fedorov (ca. 1525–1583) was the first printer in Moscow. His statue stands in front of Print Yard, what used to be the first publishing house in Moscow, between Nikolskaya Street and Teatralny Avenue, close to Lubyanka Square.

The Leichoudes Brothers

Memorial to the Leichoudes brothers (by V. M. Klykov, 2007), pioneers of higher education.

The Greek Leichoudes brothers, Ionnikios (1633–1717) and Sophronios (1653–1730), founded an academy in Moscow sometime between 1685 and 1694. Their statue is in front of the Epiphany Monastery on Bogoyavlensky Passage, opposite the exit from the Metro station Ploshchad Revolutsii. The statue was a gift from Greece to Moscow.

Fridtjof Nansen

Statue of Fridtjof Nansen (Vladimir E. Tsigal) in front of the headquarters of the Moscow Red Cross, at 6 Bolshoi Levshinsky Street.

Fridtjof Nansen (1861–1930), the Norwegian scientist and explorer, studied zoology and earned his doctorate from his research on the nervous system of lower marine organisms. He worked on behalf of prisoners of war, helped starving people in the Volga region in Russia, and assisted Armenian refugees to make it possible for them to reach safety. He was awarded the Nobel Peace Prize in 1922 for his humanitarian activities. The idea of erecting a memorial for Nansen in Moscow was raised immediately after he died, but it did not happen until seventy years later.

Architects N. A. Dygai, S. Z. Ginzburg, A. Ya. Langman, K. M. Sokolov, N. S. Streletsky, and others, lived here in the years 1940s–1960s. The plaque on the right honors Streletsky individually.

Famous architects used to live in the apartment house across the street from the Nansen statue, in Bolshoi Levshinsky Street. There is a collective memorial plaque honoring several architects and one individual plaque, specifically for Nikolai S. Streletsky (1885–1967). He was a specialist in mechanics and a designer of buildings and bridges.

House on the Embankment

House on the Embankment — House of the Government — as viewed from the Kamenny Bridge.

By about 1930, a severe shortage of living space developed in Moscow. Architect Boris M. Iofan (1891–1976) was charged with designing a big building where many of the Soviet elite would be accommodated in luxurious — luxurious, according to the then Soviet standards — apartments.

The house opened in 1931 on the Bersenevskaya Embankment of River Moscow, opposite the Kremlin; the official address is 2 Serafimovich Street. Iofan designed the building in constructivist style — a rather geometrical approach, without any sign of obsession with gigantic towers and tympanums that would characterize Soviet architecture a few years later. With over five hundred apartments and many amenities, it quickly became popular among the highest-ranking Soviet officialdom. Alas, their enthusiasm did not last long.

It has been estimated that about one third of the residents perished during Stalin's terror, the Great Purge, in the second half of the 1930s. On the site of the museum of the House, as of November 2016, there was a list of 809 names of repressed former inhabitants.[2] There are abbreviations after many of the names indicating one of the following categories: exiled, shot to death, sent to children's home (for orphaned children of the victims), died during incarceration, or committed suicide for political reasons. On the façades of the building, there are 29 memorial plaques of former residents, but only four of them honor repressed victims. According to the incomplete data of the house museum, there were at least 344 people murdered. At about the time of the sixtieth anniversary of the Great Purges, a group of concerned citizens turned to the then mayor of Moscow with a suggestion to commemorate the victims of repression with memorial plaques on the facades of buildings where the victims used to live. The mayor turned them down; he thought the facades of Moscow were already saturated with memorial plaques. There were scientists among the hundreds of victims but none among the four that did receive memorial plaques. Currently six scientists or engineers are honored with memorial plaques on the facades of the House on the Embankment.

The biochemist and academician of the Academy of Medical Sciences Olga B. Lepeshinskaya, (1871–1963) was a revolutionary under czarism and so was her husband, Panteleimon N. Lepeshinsky (1868–1944). She sided early with Lenin in his fights with rival revolutionary movements. She studied to become a medical orderly in St. Petersburg. When her husband was exiled to Eastern Russia, she accompanied him. Later, she continued her studies at the medical schools in Lausanne and Moscow, and graduated in 1915. She taught in Tashkent and in Moscow, and from 1949 she worked at the Institute of

[2] http://www.museumdom.narod.ru.

Experimental Biology — today it is the Koltsov Institute of Developmental Biology. She was head of the section of development of living matter. She and her husband lived in the House of the Government from the time it opened to the end of their days according to their memorial tablet.

Lepeshinskaya proposed a theory about regeneration of cells from structure-less "living matter." This, in essence, was the negation of cell biology that by then had prospered. Her theory was unsubstantiated, but Lysenko supported it. For years, professors of medicine were obliged to refer to Lepeshinskaya's teachings in their lectures. In 1950, a joint special commission of the Academy of Sciences and the Academy of Medical Sciences discussed her work. The commission, chaired by A. I. Oparin, fully supported Lepeshinskaya's theory. I. L. Rapoport (Chapter 1) and other renowned scientists criticized her claims. In the same year, Lepeshinskaya received the Stalin Prize whereas her critics were declared to be against Marxist ideology.

Memorial tablets of two physicians: Nikolai N. Blokhin (*left*) lived in this House from 1993 and Vladimir I. Burakovsky (*right*) from 1971, both to the end of their lives.

Nikolai N. Blokhin (1912–1993) graduated from the medical school of Lobachevsky University in Nizhny Novgorod (at the time, Gorkovsky Medical Institute). He stayed on at the university, organized regenerative surgery, and served as rector. When he moved to Moscow, he was charged with developing what is today the Blokhin Research Center of Oncology (23 Kashirskoe Highway). In two periods, he was President of the Academy of Medical Sciences for a total of 18 years. There has been considerable criticism for

Blokhin slavishly accommodating the authorities when they persecuted scientists for non-existing crimes. Blokhin himself was active in persecuting Roskin and Klyucheva (see Chapter 5), whereas he had no qualms about experimenting on humans.

Vladimir I. Burakovsky (1922–1994) studied medicine in Tbilisi, Georgia, and did postgraduate studies at the Leningrad Military Medical Academy. He moved to Moscow and first worked at the Vishnevsky Institute of Surgery, then at what is today the Bakulev Institute of Cardiovascular Surgery. From 1964, he was the director of the Bakulev Institute. His specialty was the congenital heart disorders of children and he introduced new approaches in pediatric surgery. He did research on the pathological consequences of people working under high stress.

Memorial plaques of Artem I. Mikoyan (*left*) and Petr P. Shirshov (*right*).

Artem I. Mikoyan (1905–1970), aircraft constructor, was twice Hero of Socialist Labor and brother of the top political operator Anastas I. Mikoyan. He lived in this House between 1943 and 1963. Petr P. Shirshov (1905–1953) explorer and state official lived here from 1939 to the end of his life (more about him in Chapter 7). Nikolai V. Tsitsin (see above) lived in the House from 1938 till the end of his life. The future writer, Yury V. Trifonov (1925–1981) lived in this House as a child. His father was repressed in 1937 and his mother in 1938. They were rehabilitated in 1955. Trifonov published a novel, *House on the Embankment*, which generated renewed interest to its story.

Along Tverskaya Street

7 Tverskaya Street

7 Tverskaya Street, in fall 2017, the building of the Central Telegraph opened in 1927 with the Globe on its façade. Tverskaya Street was Gorky Street in the Soviet period.

The Central Telegraph was the first structure in this most important radial street in central Moscow. On the façade of the corner part of this building there is the hammer and sickle, but not yet merging to symbolize the alliance of workers and peasants. This was the first variant of this crucial motif of the Soviet coat of arms. The building and its function were much more important in the early days of emerging technologies than today when telegraphing and telecommunications need not be centralized in one building. Nonetheless, there is great historical importance to this building. Many of the great announcements were made from this location.

Postage stamp and First Day Cover for Ashot L. Badalov and depicting the Central Telegraph.

Ashot L. Badalov's (1915–2011) life was linked with Central Telegraph. He graduated as radio engineer from the Institute of Electrical Engineering in Tbilisi in 1937. During the fateful battle of Stalingrad he was in charge of the Stalingrad radio center. From 1943, he worked in Moscow and rose from engineer to deputy minister whose responsibilities included the assignment and distribution of radio frequencies. Badalov contributed to the modernization of radio and television broadcasting in the Soviet Union and Russia.

9 Tverskaya Street

9 Tverskaya Street.

9 Tverskaya Street (designed by A. F. Zhukov) was built in 1949. The granite blocks in its foundation were the ones from which Hitler planned to construct a victory memorial of Nazi Germany over the Soviet Union. A number of noted people lived in this house, among them, the Italian physicist Bruno Pontecorvo, following his flight in 1950 to the Soviet Union. Several of the famous residents are remembered by memorial plaques. Of the over twenty memorial plaques seven honor scientists and engineers.

Left: Ivan I. Artobolevsky, (1905–1977) scientist, engineer, and trade union leader (Chapter 1) lived here between 1950 and 1975. *Right*: Andrei A. Bochvar, (1902–1984), metallurgist (Chapter 4) lived here from 1950.

Left: Yuly Khariton, who as the scientific director of Arzamas-16 lived there, had an apartment here between 1950 and 1984 (this plaque is nearly identical with the one on the facade of the Semenov Institute, Chapter 1). *Right*: Vasily Nemchinov, economist, mathematician, statistician lived here from 1950.

Yuly Khariton did not have much use of this Tverskaya Street apartment; he preferred staying at Arzamas-16. Even when his wife died in 1977, and he never married again, he did not return to Moscow where he had extended family. Even when he grew older and needed help in his everyday life, instead of returning to Moscow, members of his family stayed with him for extended periods of time.

We have already met with the economist and mathematician Vasily S. Nemchinov's (1894–1964) name in Chapter 6. He was the rector of the Timiryazev Academy at the time of the infamous August 1948 meeting of the Academy of Agricultural Sciences. He made brave attempts to save the science of genetics, but was defeated and fired from his post at the Timiryazev Academy. He and his career survived, possibly because he was not an agrarian or a biologist and represented no direct threat to Lysenko. Nemchinov pioneered the application of mathematics in economics, something that had been anathema to Stalin. Nemchinov in 1958 organized a research laboratory and in 1963 a whole research institute, what is today the Central Economics and Mathematics Institute of the Russian Academy of Sciences. It has a Mobius band on its façade (see above). The Institute helped to gain the acceptance of modern computers as part of the application of mathematical methods in regulating and planning the national economy.

Left: Konstantin Skryabin (Chapters 1 and 5), lived here from 1950. *Middle*: Nikolai V. Turbin biologist and geneticist lived here from 1971. *Right*: The organic chemist Nikolai D. Zelinsky (Chapter 1) also lived at this address.

Nikolai V. Turbin (1912–1998) was a professor at Leningrad State University, then, as member of the Belorussian Academy of Sciences (BAS), he was the director of biological institutes in Belarus. He had a controversial history; he had been a Lysenko follower and had become a Lysenko critic. In 1947, Turbin gave a devastating critique of a fellow professor, Anton R. Zhebrak (1901–1965). Zhebrak had published an article in the American *Science* journal in which he presented Soviet science in a better light than it really was. He implied that Lysenko's unscientific domination is limited to agriculture and the science of biology can still develop free of Lysenko's influence. This was too much for Soviet officialdom and Turbin expressed its displeasure by accusing Zhebrak of groveling to the West. Zhebrak was indeed a brave man, who was risking his life by trying to preserve the memories of N. I. Vavilov and his perished associates. This was in 1947.

In 1948, at the infamous August session of the Lenin Academy of Agricultural Sciences Turbin gave a talk in full support of Lysenkoism and Iosif Rapoport interrupted him by screaming "Obscurantists!" meaning people who hinder truth and progress.[3] When in 1948 serious biologists lost their jobs at Leningrad University, the head of the Department of Genetics, Mikhail Lobashov, was replaced by Turbin. By 1952 Turbin had apparently changed his mind and published an article in the *Botanical Journal* in which he called Lysenko's claims unsubstantiated. The *Botanical Journal* published papers critical of Lysenkoism, but it offered Lysenko its forum as well. The Communist Party was not satisfied; it intervened, and the whole editorial board of the journal was dismissed. Although Turbin took such steps to mitigate his prior behavior, the noted science historian Simon E. Shnoll named Turbin one of the eight most disgraceful scientists of the Lysenko affair.[4]

11 Tverskaya Street

11 Tverskaya Street, the Ministry of Science and Higher Education and the Ministry of Education of the Russian Federation.

[3] Vadim J. Birstein, *The Perversion of Knowledge: The True Story of Soviet Science* (Cambridge, MA: Westview Press, 2001), p. 270.

[4] Simon E. Shnoll, *Geroi, zlodei, konformisti rossiiskoi nauki* (Heroes, villains, conformists of Russian science, third edition, Moscow: URSS, 2009), p. 548.

In 2018, the Ministry of Education and Science was split into the Ministry of Science and Higher Education and the Ministry of Education.

House of Scientists

The Central House of Scientists at 16 Prechistenka Street is a club of scientists for relaxation, meetings with politicians and artists, and science popularization. The original building was designated to be the House of Scientists in 1922. Famous scientists, Russian and international, give presentations about their science to a broad audience. We mention only one. On June 3, 2008, James D. Watson gave a public lecture in this House with the title "DNA and the Brain: In search of the genes of psychiatric illnesses." Watson's lecture drew a huge crowd, especially of young people. Back in the United States his reputation had just been damaged by his unfortunate interview in which he talked about the gloomy future of Africa because of the allegedly low IQ of its people. In Moscow, his popularity was unscathed.

Watson has given other lectures in Moscow, always in front of large audiences. On June 15, 2010, the title of his presentation was "The Rules of Success" at the Faculty of Biology of Moscow University. Five years later, on June 17, 2015, Watson gave another lecture in Moscow, entitled "From the Structure of DNA to Curing Cancer." He delivered this lecture at the headquarters of the Academy of Sciences and the occasion was rather peculiar. Some months earlier, Watson had his Nobel gold medal auctioned. Alisher Usmanov, a Russian billionaire entrepreneur, bought it for over four million dollars, but he wanted the scientist who had earned the medal to own it. On the day of the lecture, Watson met with Usmanov in Moscow at the headquarters of the Science Academy. After the lecture, Vladimir E. Fortov, the then president of the Academy, handed the Nobel medal to Watson on Usmanov's behalf.

Kremlin Wall

Portions of the Red Square with the Kremlin Wall.

The previous chapter introduced us to a most prestigious burial place in Moscow, the Novodeviche Cemetery, where many renowned scientists are buried. The most prestigious burial place is, of course, the Lenin Mausoleum, followed by the twelve graves behind the Mausoleum, each with a bust of the buried, among them Iosif V. Stalin's. There is then the Kremlin Wall that is the venue of internment mostly for revolutionaries and military and political leaders. Some of the scientists and technologists have risen so high in the hierarchy of officialdom that rendered them a place in the Kremlin Wall.

Grave plates of Mstislav V. Keldysh (*left*) and Yury A. Gagarin (*right*) in the Kremlin Wall.

We have counted 11 plates of people, including some cosmonauts, belonging to the scope of this book. The plates are all uniform, hence no need to present each of them, but we say a few words about those that did not figure elsewhere in our discussion. Georgy T. Dobrovolsky, (1928–1971), cosmonaut, lost his life in 1971 in the flight of the space ship Soyuz-11. Its three crew members were commander Dobrovolsky, engineer-researcher Viktor I. Patsaev (1933–1971), and engineer Vladislav M. Volkov (1935–1971). The team was returning to Earth after over 23 days in space (a record then) and a successful stay in the Space Station Solyut. Upon the separation of the return unit from the rest of the space ship, a valve stayed open, the cabin of the return unit depressurized — there were conditions of vacuum in the cabin — and the three men died of suffocation.

Yury A. Gagarin (1934–1968), cosmonaut, was the first man in space on April 12, 1961 (Chapter 4). He lost his life in an aircraft test flight.

Aleksandr P. Karpinsky (1847–1936), geologist and mineralogist, was a long-time president of the Soviet Academy of Sciences (Chapter 1).

Mstislav V. Keldysh (1911–1978), chief scientist of the space program (Chapter 4).

Vladimir M. Komarov (1927–1967), cosmonaut, lost his life when his spaceship crashed.

Sergei P. Korolev, (1906–1966), chief designer of the space program (Chapter 4).

Leonid A. Kostandov (1915–1984) was in charge of Soviet chemical industry (Chapter 4).

Igor G. Kurchatov (1903–1960), nuclear physicist, the "father of the Soviet atomic bomb" (Chapter 1).

The grave plates of Vyacheslav A. Malyshev (*left*) and Avraami P. Zavenyagin (*right*) in the Kremlin Wall.

Vyacheslav A. Malyshev (1902–1957) graduated as engineer from Bauman University in 1934. In 1939–1940 he was the minister of heavy industry and advocated the need of preparedness for the approaching war. He had high positions in the government in charge of machine building, tank production, and as deputy chairman of the Council of Ministers. He was the first head of the Ministry of Medium Machine Building — cover name for the ministry of the nuclear project. In 1953 he rushed to the site of the explosion of the first Soviet thermonuclear device and his premature death may have been the consequence of the radiation he received on this occasion.

Avraami P. Zavenyagin (1901–1956), a metallurgist engineer, was a leading figure in the Soviet nuclear program. He studied at the Moscow Mining Academy and was its vice rector even while he was its student. He served in various leading industrial positions. During the late 1930s he was director of a plant with thousands of slave laborers. From 1941 Zavenyagin was one of the top ranking members of the Ministry of the Interior and State Security. From 1943, he was in the leadership of the atomic bomb project along with L. P. Beriya, M. G. Pervukhin, V. A. Malyshev, and B. L. Vannikov.

In 1945 Zavenyagin was in charge of the operation to find and move German physicists, chemists, and technologists to the Soviet Union and organize their work in Soviet laboratories. Yu. B. Khariton and I. K. Kikoin were among the members of Zavenyagin's team. They found the German reserves of uranium ore and uranium compounds and had them transported to the Soviet Union. In 1948, there were two accidents at the first Soviet industrial nuclear reactor and it is supposed that he spent too much time in the danger zone and received a life-threatening amount of radiation. In 1949, he was on the scene after the first successful nuclear explosion in the Soviet Union. He was driven to the site of the explosion, but had to walk back when the car could not be operated in the dust. On this occasion, again, he may have received too much radiation. In 1955, he was appointed minister of Medium Machine Construction.

Other Cemeteries

Beside the Novodeviche Cemetery, a number of cemeteries in Moscow are the final resting place of famous scientists and technologists. Here we present a sampler of other cemeteries with a few personalities to indicate the significance of these venues.

The grave of Nikolai Zhukovsky at the Donskoe Cemetery. https://upload.wikimedia.org/wikipedia/commons/6/61/Tomb_in_Cemetery_in_Donskoy_Monastery_11.jpg; by "Bogdanov-62;" 01/07/17.

Donskoe Cemetery at 4 Ordzhonikidze Street is known as the resting place of many of the repressed victims of Stalinism. There are also many graves of scientists and technologists. We show here the grave of Nikolai E. Zhukovsky (Chapter 4).

Grave stone of Evgeny Zavoisky (courtesy of E. B and N. E. Zavoiskaya) at the Kuntsevskoe Cemetery.

Kuntsevskoe Cemetery at 20 Ryabinovaya Street has, among others, the grave of Evgeny K. Zavoisky, the discoverer of the electron paramagnetic resonance technique for analysis of materials structure (Chapter 1). Also, there is the grave of Trofim D. Lysenko (Chapter 1) marked by a simple black marble stone with an eastern Orthodox Cross as the only decoration on it.

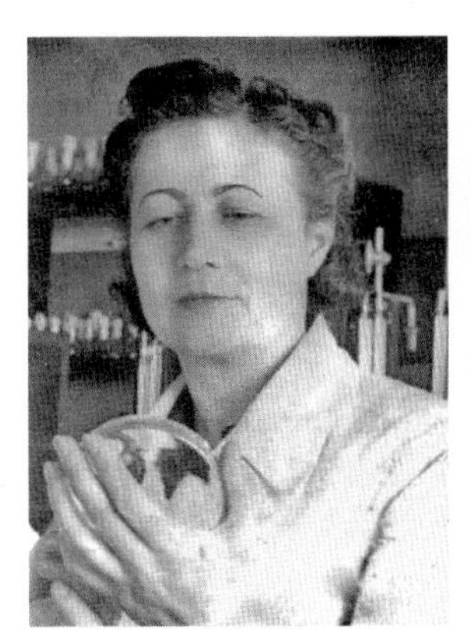

Zinaida Ermoleva in the laboratory (httpen.ww2awards.comperson69020#top; 01/17/17) and detail of her gravestone relief (http://fb.ru/article/242126/zinaida-vissarionovna-ermoleva-biografiya-i-foto; accessed 01/17/17) at the Kuzminskoe Cemetery.

Kuzminskoe Cemetery, at 19 Akademik Skryabin Street, is the resting place for Zinaida V. Ermoleva (1898–1974), a microbiologist, who was the first to produce penicillin in the Soviet Union. She did this during WWII, which had great strategic significance. She was working in parallel with the British scientists and when Howard Florey, the pioneer of penicillin, visited Moscow and compared the British and Russian samples, he expressed his admiration for Ermoleva and called her *Lady Penicillin*.

Ermoleva worked out preventive measures against cholera, which included even infecting herself in order to find the most efficient measures. She conducted life-long studies of antibiotics and headed laboratories and institutes concerned with antimicrobial materials of natural origin. During the last two decades of her life she was in charge of the microbiology department at the Central Post-Graduate Medical Institute in Moscow.

She was married for a few years to the renowned virologist Lev A. Zilber (Chapter 5). Although they had divorced, she stood by him at the time of his incarceration on trumped up charges. She fought bravely to have Zilber freed and he was in 1944. Her second husband, the microbiologist Aleksei A. Zakharov was also arrested on false charges and died in prison in 1940. Her persona served as model for the heroine of a novel, a feature film, and two TV series. The fictitious though realistic character in the novel became a role model for girls embarking on a career in microbiology.

Memorial of the victims and heroes of the Chernobyl Catastrophe. In front of the central monument there are the memorials of the first respondent firefighters (https://ru.wikipedia.org/wiki/Файл:Chernobyl_memorial.jpg; by "Anatolich1;" 01/15/17).

Mitinskoe Cemetery, Pyatnitskoe Highway, at kilometer 6 is a relatively new cemetery. The most famous person buried there is Anatoly P. Aleksandrov (Chapter 1). He was instrumental in the development of nuclear reactors, power stations, nuclear icebreakers, and nuclear submarines. He was in charge of the construction of the Chernobyl reactor. He and his wife are buried at the segment 144 in the Mitinskoe Cemetery rather than at Novodeviche Cemetery for which he, thrice Hero of Socialist Labor, would have been entitled, and what many sources have, mistakenly, communicated. Moscow's Chernobyl Memorial was also erected at this Cemetery in 2011.

The Chernobyl Memorial in the Mitinskoe Cemetery was unveiled on April 26, 2011. It honors the victims of the Chernobyl catastrophe that happened during the night of April 25 to 26, 1986. There was an explosion at the Chernobyl Nuclear Power Station signaling the beginning of the most terrible industrial disaster. There were immediate fatalities, including those working the night shift and the first respondent firefighters. It is estimated that altogether some 800,000 firefighters, emergency medical personnel, engineering troops, specialists of chemical protection, drivers, helicopter teams, and countless others in addition to the inhabitants of the region were exposed to unknown amounts of radiation.

By quirky coincidence, Anatoly P. Aleksandrov's grave, together with his wife, is not far from the Chernobyl Memorial, though in a different section. Nothing on his gravestone indicates his high positions in the Soviet scientific establishment and his three stars of the Hero of Socialist Labor, neither his intimate involvement with the nuclear projects, including the design of nuclear power plants.

Tombstone of Vladimir Gerie (Guerrier), founder of the Moscow Higher Courses for Women, at the Pyatnitskoe Cemetery (https://commons.wikimedia.org/wiki/File:Piatnitskoe_Cemetery_280811_Guerrier_Tomb.jpg; by "Vladimir OKC;" 01/15/17).

Pyatnitskoe Cemetery, at 3 Droboliteiny Street (this is a short connecting road from Mir Avenue to the cemetery). Here we single out Vladimir I. Gerie (Guerrier, 1837–1919) who was a professor of history at Moscow University. He founded the Moscow Higher Courses for Women. He was unhappy about the ministerial order in 1863 to reject women for enrollment at Moscow University. By then he had gathered excellent experience about the dedication and preparedness of women students who came to be examined for becoming private tutors. He prepared the initial set of rules for the new institution, and in 1872, the new organization opened. Gerie stayed director until 1888 and was director again between 1900 and 1905.

Troyekurovskoe Cemetery, at 24 Ryabinovaya Street. From among the scientists buried here, we mention Sergei A. Khristianovich and Andrei D. Mirzabekov (both figure in Chapter 1).

Vagankovskoe Cemetery, at 15 Sergei Makeev Street. From among the scientists buried here, we mention Vladimir P. Demikhov (Chapter 5) and Vsevolod M. Klechkovsky (Chapter 6).

The grave stone of Lev Altshuler and his wife, Mariya Speranskaya at the Vostryakovskoe Cemetery (courtesy of Boris Altshuler).

Vostryakovskoe Cemetery, at 47 Ozernaya Street. The gravestone of Andrei D. Sakharov in the Vostryakovskoe Cemetery is shown in Chapter 1. Lev V. Altshuler (1913–2003) was one of his long-time physicist colleagues at the secret Arzamas-16 nuclear laboratory. Altshuler graduated in physics from Moscow University in 1936. He fought in WWII, but was soon recalled from the front and engaged in defense-related research. He was one of the leading associates at Arzamas-16 between 1946 and 1969. His principal work was in the investigation of materials properties under high pressure and high temperature conditions and under the impact of shock waves. He was the founder of the Russian school of research of dynamic processes. His job, even his life, was threatened in 1951, during the anti-Semitic wave of terror during Stalin's last years. A committee of secret police agents descended upon Arzamas-16 to cleanse it of Jewish personnel. Altshuler's situation was especially precarious because of his views concerning modern music and biology that diverged from the party line. Today, this sounds absurd, but at the time he could have ended up in a slave labor camp or worse. The scientific head of Arzamas-16 advised Altshuler to stay home for a while and his becoming invisible may have saved him.

The grave stones of Nikolai Luzin (*left*) and two members of his school of mathematics "Luzitania," Nina Bari (*middle*) and Aleksei Lyapunov (*right*), at the Vvedenskoe Cemetery (all three images, courtesy of Nikolai Andreev).

Vvedenskoe Cemetery, at Nalichnaya Street 1. Nikolai N. Luzin (Chapter 2) and a number of other mathematicians are buried at this cemetery. Nina K. Bari (1901–1961) was one of the first female students of mathematics at Moscow University, she was one of Luzin's pupils, and eventually a member of Luzin's circle, "Luzitania," while it existed. Bari herself became a professor of mathematics at Moscow University and her research area was the so-called trigonometric series. She was killed when she fell in front of a metro train. Her mathematician husband Viktor V. Nemytsky is buried next to her grave.

Aleksei A. Lyapunov (1911–1973) was a nephew of the world-renowned mathematician Aleksandr M. Lyapunov. Aleksei studied mathematics at Moscow University, but left it well before graduation. Later he completed his studies on his own as Luzin's student and became a member of "Luzitania." Lyapunov worked at the Steklov Institute, fought in WWII, and in 1953 he joined Mstislav V. Keldish when Keldysh established his new institute of applied mathematics. Lyapunov was one of the founders of cybernetics in the Soviet Union at the time when this area of science was considered anti-Soviet. From 1955, the situation eased, and Lyapunov's work became part of the main stream. During the last decade of his life, Lyapunov held leading positions in the Siberian Branch of the Academy of Sciences and at Novosibirsk University. One of his recognitions was the "Computer Pioneer" medal.

From among the graves of other scientists we mention here those of the experimental biologist Nikolai K. Koltsov and his wife, Mariya P. Sadovnikova-Koltsova (Chapter 1) and the Nobel laureate theoretical physicist Ilya M. Frank (Chapter 1).

Select Bibliography

Alferov, Zh, Editor-in-Chief, *Ioffe Institute 1918–1998: Development and Research Activities* (St. Petersburg: Ioffe Physico-Technical Institute, 1998).

Alpert, Yakov, *Making Waves: Stories from My Life* (New Haven, CT: Yale University Press, 2000).

Altshuler, B. L., Ed., *Andrei Sakharov: Facets of a Life* (Gif-sur-Yvette, France: Edition Frontiers, 1991).

Andreev, A. F., Ed., *Kapitza Tamm Semenov: V ocherkakh i pis'makh* (Kapitza, Tamm, Semenov: In Sketches and Letters; Moskva: Vagrius Priroda, 1998).

Applebaum, Anne, *Gulag: A History* (New York: Anchor Books, 2004).

Atomnii Proekt SSSR I 1938–1945 (Moscow: Nauka-Fizmatlit, 1999); *Atomnii Proekt SSSR II 1945–1954* (Moscow-Sarov: Nauka-Fizmatlit, 1999).

Badash, Lawrence, *Kapitza, Rutherford, and the Kremlin* (New Haven and London: Yale University Press, 1985).

Baev, A. A., Ed., *Vospominaniya o V. A. Engelhardte* (Remembering V. A. Engelhardt, Moscow: Nauka, 1989).

Baggott, Jim, *Atomic: The First War of Physics and the Secret History of the Atom Bomb: 1939–1949* (London: Icon Books, 2009).

Birstein, Vadim J., *The Perversion of Knowledge: The True Story of Soviet Science* (Cambridge, MA: Westview Press, 2001).

Blokh, A. M., *Sovetskii Soyuz v interyere nobelevskikh premii: Fakti. Dokumenti. Razmyshleniya. Kommentarii.* Second Edition (The Soviet Union in the interior of the Nobel Prizes. Facts; Documents; Reflections; Commentaries, Moscow: Fizmatlit, 2005). English translation: Abram M. Blokh, *Soviet Union in the Context of the Nobel Prize* (Singapore, etc.: World Scientific, 2018).

Blokh, A. M., *Nobelevskaya premiya — populyarno obo vsem* (Nobel Prize — popularly about everything; Moscow: BuKoc, 2008).

Boag, J. W., P. E. Rubinin, and D. Shoenberg, Compilers and Editors, *Kapitza in Cambridge and Moscow: Life and letters of a Russian Physicist* (Amsterdam: North-Holland, 1990).

Bongard-Levin, G. M. and Zakharov, V. E., Eds., *Rossiiskaya nauchnaya emigratsiya: Dvadtsat portretov* (Scientific Emigration from Rossiya: Twenty Portraits; Moscow: URSS, 2001).

Bonner, Elena, *Alone Together* (New York: Vintage Books, 1988).

Borvendég, Zsuzsanna, and Palasik, Mária, *Vadhajtások: A sztálini természetátalakítási terv átültetése Magyarországon 1948–1956* (Wildings: The adaptation of the

Stalin plan of remaking Nature in Hungary 1948–1956; Budapest: Napvilág Kiadó, 2015).

Bové, Joseph, *Памятники архитектуры в дореволюционной России* (Moscow: Terra, 2002).

Chleny Rossiiskoi akademii nauk v Matematicheskom institute im. V. A. Steklova RAN (Members of the Russian Academy of Sciences at the Steklov Mathematical Institute of RAS; Moscow: Matematicheskoi Institut im. V. A. Steklova RAN, 2009).

Devyatkova, Lyudmila, *Akademik R. V. Khokhlov — rektor Moskovskogo Universiteta* (Academician R. V. Khokhlov — rector of Moscow University; Moscow: Izdatelstvo Moskovskogo Universiteta, 2005).

Fedin, E. I., *Filin na razvalinakh* (Eagle-owl over the ruins; St. Petersburg, 2000).

Fedin, Erlen, *Izbrannoe* (Selected Works, Krasnoyarsk: Polikom, 2008).

Feifer, Gregory, *Russians: The People behind the Power* (New York and Boston: Twelve, 2015).

Feinberg, E. L., *Physicists: Epoch and Personalities* (New Jersey, London, Singapore: World Scientific, 2011).

Felshtinskii, Yu. G., *Razgovori s Bukharinym* (Conversations with Bukharin; New York: Telex, 1991).

Gamow, George, *My World Line: An Informal Autobiography* (New York: Viking Press, 1970).

Ginzburg, Vitaly L., *About Science, Myself and Others* (Bristol and Philadelphia: Institute of Physics Publishing, 2005).

Ginzburg, Vitaly L., *O fizike i astrofizike* (About Physics and Astrophysics, third revised edition, Moscow: Buro Quantum, 1995).

Gol'danskii, Vitalii I., *Essays of a Soviet Scientist: A revealing portrait of a life in science and politics* (Woodbury, NY: American Institute of Physics, 1997).

Goldansky, V. I., A. Yu. Semenov, and M. B. Chernenko, Eds., *Yulii Borisovich Khariton: Put' dlinoyu v vek* (Century-long journey; Moscow: Nauka, 2005).

Gorelik, Gennady, with Antonina W. Bouis, *The World of Andrei Sakharov: A Russian Physicist's Path to Freedom* (New York: Oxford University Press, 2005).

Gorobets, Boris S., *Krug Landau i Lifshitsa* (Landau's and Lifshits's Circle, Moscow: URSS, 2008).

Gorobets, Boris S., *Krug Landau: Fizika voini i mira* (Landau's Circle: Physics of War and Peace; Moscow: URSS, 2009).

Gorobets, Boris S., *Krug Landau: Zhizn geniya* (Landau's Circle: The life of a genius, 2nd corrected and augmented edition, Moscow: URSS, 2008).

Gorobets, Boris S., *Sekretnie fiziki iz atomnogo proekta SSSR: Semya Leipunskikh* (Classified physicists from the atomic project of the Soviet Union: The Leipunskii family, Moscow: URSS, second edition, 2009).

Gorobets, Boris S., *Sovetskie fiziki shutyat ... khotya bivalo ne do shutok* (Soviet physicists joking... although it was not joyful; Moscow: URSS, 2010).

Gorobets, Boris S., Yaderny revansh Sovetskogo Soyuza: Ob istorii atomnogo proekta (Nuclear revenge of the Soviet Union: About the history of the atomic project; Moscow: URSS, 2013).

Gorobets, Boris S., Yaderny revansh Sovetskogo Soyuza: Sudby Geroev, dvazhdy Geroev, trizhdy Geroev atomnoi epopei (Nuclear revenge of the Soviet Union: Fates of Heroes, double Heroes, triple Heroes of the atomic epopee, Moscow: URSS, 2013).

Goudsmit, Samuel A., *Alsos* (New York: Henry Schuman, 1947).

Graham, Loren R., *Science, Philosophy, and Human Behavior in the Soviet Union* (New York: Columbia University Press, 1987).

Graham, Loren R., *Science in Russia and the Soviet Union: A Short History* (Cambridge, UK: Cambridge University Press, 1993).

Graham, Loren, *Lonely Ideas: Can Russia Compete?* (Cambridge, MA, and London, England: The MIT Press, 2013).

Hargittai, Balazs, and István Hargittai, *Candid Science V: Conversations with Famous Scientists* (London: Imperial College Press, 2005).

Hargittai, Istvan, *Candid Science: Conversations with Famous Chemists* (edited by Magdolna Hargittai, London: Imperial College Press, 2000).

Hargittai, Istvan, *Candid Science II: Conversations with Famous Biomedical Scientists* (edited by Magdolna Hargittai, London: Imperial College Press, 2002).

Hargittai, Istvan, *Candid Science III: More Conversations with Famous Chemists* (edited by Magdolna Hargittai, London: Imperial College Press, 2003).

Hargittai, Istvan, *Our Lives: Encounters of a Scientist* (Budapest: Akadémiai Kiadó, 2004).

Hargittai, Istvan, *Martians of Science: Five Physicists Who Changed the Twentieth Century* (New York: Oxford University Press, 2006).

Hargittai, Istvan, *Judging Edward Teller: A Closer Look at One of the Most Influential Scientists of the Twentieth Century* (Amherst, NY: Prometheus Books, 2010).

Hargittai, Istvan, *Drive and Curiosity: What Fuels the Passion for Science* (Amherst, NY: Prometheus, 2011).

Hargittai, Istvan, *Buried Glory: Portraits of Soviet Scientists* (New York: Oxford University Press, 2013).

Hargittai, Istvan, and Magdolna Hargittai, *In Our own Image: Personal Symmetry in Discovery* (New York: Kluwer Academic/Plenum Publishers, 2000).

Hargittai, Istvan, and Magdolna Hargittai, *Candid Science VI: More Conversations with Famous Scientists* (London: Imperial College Press, 2006).

Hargittai, Magdolna, and Istvan Hargittai, *Candid Science IV: Conversations with Famous Physicists* (London: Imperial College Press, 2004).

Holloway, David, *Stalin and the Bomb: The Soviet Union and Atomic Energy 1939–1956* (New Haven & London: Yale University Press, 1994).

Hyde, H. Montgomery, *Stalin: The History of a Dictator* (New York: Popular Library, 1971).

Ings, Simon, *Stalin and the Scientists: A History of Triumph and Tragedy 1905–1953* (New York: Atlantic Monthly Press, 2016).

Ivanov, V. T., Ed., *Yurii Anatolevich Ovchinnikov: Zhizn i nauchnaya deyatelnost* (Yurii Anatolevich Ovchinnikov: Life and Scientific Oeuvre; Moscow: Nauka, 1991).

Josephson, Paul R., *Red Atom: Russia's Nuclear Power Program from Stalin to Today* (New York: W. H. Freeman and Co., 2000).

Josephson, Paul R., *Lenin's Laureate: Zhores Alferov's Life in Communist Science* (Cambridge, MA, London: MIT Press, 2010).

Kabachnik, M. I., Ed., *Aleksandr Nikolaevich Nesmeyanov: Uchonii i Chelovek* (Aleksandr Nikolaevich Nesmeyanov: Scientist and Human Being; Moscow: Nauka, 1988).

Kapitza, P. L., *Experiment, Theory, Practice* (Dordrecht, Boston, London: D. Reidel, 1980).

Khalatnikov, I. M., *Dau, Kentavr i drugie (Top nonsecret)* (Dau, Centaur and Others (Top non-secret); Moscow: Fizmatlit, 2008).

Khalatnikov, I. M., Ed., *Vospominaniya o L. D. Landau* (Reminiscences about L. D. Landau; Moscow: Nauka, 1988).

Khrushchev, N. S., *Khrushchev Remembers: The Last Testament* (Little Brown & Co., 1974).

Kisselev, Lev L., "Half a Century Later and, Still, I'm Not Disenchanted with Science." In V. P. Shulachev and G. Semenza, Eds., *Stories of Success — Personal Recollections. XI. Comprehensive Biochemistry*, Vol. 46. (Elsevier, 2008).

Kiselev, L. L. and Levina, E. S., *Lev Aleksandrovich Zilber (1894–1966) Zhizn v nauke* (Lev Aleksandrovich Zilber (1894–1966): A life in science; Moscow: Nauka 2004).

Knight, Amy, *Beria: Stalin's First Lieutenant* (Princeton, NJ: Princeton University Press, 1993).

Kojevnikov, Alexei B., *Stalin's Great Science: The Times and Adventures of Soviet Physicists* (London: Imperial College Press, 2004).

Krylov, N. L., Klyuzhev, V. M., and Maksimov, I. B., Eds., *Pervy gospital i voennaya meditsina Rossii: Tom I, Stanovlenie voennoi meditsiny Rossii. Knigi 1 i 2* (The first hospital and military medicine of Russia, Volume I, The emergence of Russian military medicine. Books 1 and 2; Moscow: Eko-Press, 2010).

Kumanev, V. A., *Tragicheskie Sudbi: Repressirovannie uchonie Akademii nauk SSSR* (Tragic Fates: Suppressed scientists of the Soviet Academy of Sciences; Moscow: Nauka, 1995).

Kuznetsova, N. I., Ed., *Chelovek, kotorii ne umel bit ravnodyshnim: Yurii Timofeevich Struchkov v nauke i zhizhni* (The man who was unable to be indifferent: Yurii Timofeevich Struchkov in science and in life; Moscow: Russian Academy of Sciences, 2005).

Leonova, E. B. (compiler), *A. I. Kitaigorodskii: Uchonii, Uchitel, Drug* (A. I. Kitaigorodskii: Scientist, Teacher, Friend; Moscow: Moskvovedenie, 2011).

Levinshtein, Michael, *The Spirit of Russian Science* (New Jersey, London, Singapore: World Scientific, 2002).

Lobikov, E. A., *Sovremennaya Fizika i Atomnii Proekt* (Modern Physics and Atomic Project; Moscow and Izhevsk: Institut Komputernikh Issledovanii, 2002).

Lunin, V. V., Ed., *Khimichesky Fakultet MGU* (Faculty of Chemistry of Moscow State University; Moscow: Terra-Kalender, 2005).

Lunin, V. V., Ed., *Zhenshchiny-Khimiki* (Women-Chemists; Moscow: Yanus-K, 2013).

Mackay, Alan L., *A Dictionary of Scientific Quotations* (Bristol: Adam Hilger, 1991).

Montefiori, Simon Sebag, *Stalin: The Court of the Red Tsar* (New York: Alfred A. Knopf, 2004).

Moss, Walter G., *A History of Russia, Volume II: Since 1855* (New York: McGraw-Hill Co., 1997).

National Air and Space Museum (Third Edition, Washington, DC: Smithsonian Books, 2009).

Nauchnye shkoly Moskovskogo Gosudarstvennogo Tekhnicheskogo Universiteta imeni N. E. Baumana (The scientific schools of the Bauman Moscow State University of Technology; Moscow: Izdatelstvo MGTU im. N. E. Baumana, 1995).

Nesmeyanov, A. N., *Na kachelyakh XX veka* (Sitting on the swings of the twentieth century, Moscow: Nauka, 1999).

Nobel Lectures: Physics 1971–1980 (Singapore: World Scientific, 1992).

Ozernyuk, N. D., *Nauchnaya shkola N. K. Koltsova: Ucheniki i soratniki* (N. K. Koltsov's scientific school: Pupils and colleagues; Moscow: Institut biologii razvitiya im. N. K. Koltsova RAN, 2012).

Ozkan, Svetlana, *Novodevichy Necropolis in Moscow* (Moscow: Ritual, 2007).

Paloczi-Horvath, George, *The Facts Rebel: The Future of Russia and the West* (London: Secker & Warburg, 1964).

Pechenkin, Alexander, *Leonid Isaakovich Mandelstam: Research, Teaching, Life* (Heidelberg, etc.: Springer 2014).

Pollock, Ethan, *Stalin and the Soviet Science Wars* (Princeton, NJ: Princeton University Press, 2006).

Popovsky, Mark, *Science in Chains: The Crisis of Science and Scientists in the Soviet Union Today* (London: Collins and Harvill Press, 1980).

Popovsky, Mark, *The Vavilov Affair* (Hamden, CT: Archon Books, 1984).

Popovsky, Mark, *Upravlyaemaya nauka* (Regulated Science, Free electronic library, royallib.ru [July 23, 2017]).

Pringle, Peter, *The Murder of Nikolai Vavilov: The Story of Stalin's Persecution of One of the Great Scientists of the Twentieth Century* (New York: Simon & Schuster, 2008).

Ramensky, E. V., *Nikolai Koltsov: Biolog, obognavshij vremya* (Nikolai Koltsov: A biologist ahead of his age; Moscow: Nauka, 2012).

Redlich, Shimon, *War, Holocaust and Stalinism: A Documented History of the Jewish Anti-Fascist Committee in the USSR* (Luxembourg: Harwood Academic Publishers, 1995).

Roll-Hansen, Nils, *The Lysenko Effect: The Politics of Science* (Amherst, NY: Humanity Books, 2005).

Sagdeev, Roald Z., *The Making of a Soviet Scientist: My Adventures in Nuclear Fusion and Space from Stalin to Star Wars* (New York: John Wiley & Sons, 1994).

Sakharov, Andrei, *Memoirs* (translated by Richard Lourie; New York: Alfred A. Knopf, 1990).

Sakharov, Andrei, *Moscow and Beyond 1986–1989* (New York: Alfred A. Knopf, 1991).

Sardanashvili, Genaddii A., *Dmitrii Ivanenko — Superzvezda sovetskoi fiziki: Nenapisennie memuari* (Dmitrii Ivanenko — Superstar of Soviet Physics: Unwritten memoirs; Moscow: URSS, 2010).

Selected works of Yakov Borisovich Zeldovich. Volume I. Chemical Physics and Hydrodynamics (Princeton, NJ: Princeton University Press, 1992).

Selected works of Yakov Borisovich Zeldovich. Volume II. Particles, Nuclei, and the Universe (Princeton, NJ; Princeton University Press, 1993).

Semenov, Nikolai N., *Tsepnie Reaktsii* (Chain Reactions, Leningrad: Goskhimizdat, 1934; in English translation, Oxford: Oxford University Press, 1935).

Shikheeva-Gaister, Inna, *Deti vragov naroda: Semeinaya khronika vremen kulta lichnosti 1925–1953* (Children of the enemies of people: Family chronicle from the time of the cult of personality 1925–1953; Moscow: Vozvrashchenie, 2012).

Shilov, A. E., Ed., *Vospominaniya ob akademike Nikolae Nikolaeviche Semenove* (Reminiscences about academician Nikolai Nikolaevich Semenov; Moscow: Nauka, 1993).

Shilov, A. E. *et al.*, Eds., *Nikolai Nikolaevich Semenov* (Moscow: Palma Press, 2006).

Shmidt, T., *Vremya, lyudi i sudby "Dom na naberezhnoi"* (Time, people and fates "House on the Embankment"; Moscow: Vozvrashchenie, 2015).

Shnoll, Simon E., *Geroi, zlodei, konformisti rossiiskoi nauki* (Heroes, villains, conformists of Russian science, third edition, Moscow: URSS, 2009).

Shubnikov, A. V. and V. A. Koptsik, *Symmetry in Science and Art* (New York and London: Plenum Press, 1974). This is a later expanded version of Shubnikov's original book in Russian.

Sidur, Mikhail V., *Vadim Sidur* (Moscow: Moskovsky Gosudarstvenny Muzei Vadima Sidura, 2004).

Smolegovsky, A. M., *I. I. Kitaigorodsky i ego trudi v oblasti khimii i khimicheskoi tekhnologii stekla, keramiki i sitallov* (I. I. Kitaigorodsky and his works in chemistry and chemical technology of glass, ceramics, and sitals; Perm, Russia: Bazaltovie Tekhnologii, 2005).

Sonin, A. S., *"Fizicheskii idealism" Istoriya odnoi ideologicheskoi kampanii* ("Physical idealism" The history of an ideological campaign; Moscow: Fiziko-Matematicheskaya Literatura, 1994).

Sostoyanie teorii khimicheskogo stroeniya v organicheskoi khimii (State of the Theory of Chemical Structure in Organic Chemistry: All-Union Conference 11–14 June, 1951, stenographic minutes; Moscow: Publishing House of the Academy of Sciences of the USSR, 1952).

Spufford, Francis, *Red Plenty* (London: Faber and Faber, 2010).

Sukhomlinov, Andrei, *Kto Vi, Lavrentii Beria? Neizvestnie stranitsi uglovogo dela* (Who Are You, Lavrentii Beria? Unknown pages of a criminal case; Moscow: Detektiv-Press, 2003).

Sunyaev, Rashid A., Ed., *Zeldovich: Reminiscences* (Boca Raton, LA: CRC Press, 2005).

Weiner, Douglas R., *A Little Corner of Freedom: Russian Nature Protection from Stalin to Gorbachev* (Berkeley, CA: University of California Press, 1999).

Zeldovich, Yakov B., *My Universe: Selected Reviews* (London: Routledge, 1992).

Index*

1812 Patriotic War 206, 211, 236
1896 Exhibition (Nizhny Novgorod) 133, 246
1911 protest 6, 48, 70, 88, 97, 98, 219, 220, 238, 240, 278, 292, 325
1917 revolutions 62, 65, 70, 77, 171, 181, 186, 197, 232, 267, 283
1948 session of the Academy of Agricultural Sciences 38, 63, 252, 290, 339, 340

A
Abich, Herman von 125
Abramov, Pavel V. 276
Abrikosov, Aleksei A. 14, 31, 88, 143, 184, 291
Abrikosov, Aleksei I. 184, 211, 290, 291
Abrikosova (née Vulf), Fanni D. 184, 290, 291
Abrosimov, Pavel V. 277
academy elections 4, 18, 21, 267
Academy of Agricultural Sciences 3, 38, 57, 60, 252, 290, 339, 340
Academy of Medical Sciences 3, 99, 198, 199, 207, 214, 333, 334
Academy of Medical Surgery 187
Academy of Military Surgery 190
Aeroflot 274
Akulov, Nikolai 38
Alabyan, Karo S. 295
Al-Biruni, Abu-Reikhan 110
Alekhin, Vasily V. 122, 286
Aleksandrov, Aleksandr P. 166, 277
Aleksandrov, Anatoly P. 19, 32, 346, 347
Aleksandrov, Pavel S. 85, 86
Aleksandrovsky Garden 80n
Alekseev Psychiatric Clinical Hospital 188
Alekseev, Nikolai A. 188
Alekseevskaya Eye Hospital →Helmholtz Institute
Alekseevsky, E. E. 31
Alexander II 191
Alferov, Zhores I. 311
Alikhanov Institute of Theoretical and Experimental Physics 262
Alikhanov, Abram I. 16, 91, 261, 262
All-Russia Exhibition Center (VDNKh) 326
All-Union Lenin Academy of Agricultural Sciences →Academy of Agricultural Sciences
Altshuler, Lev V. 348, 349
Amalitsky, Viktor P. 324
Ambartsumyan, Viktor A. 319, 320
American Association of Surgeons 203
Amundsen, Roald 103
Andreev Acoustics Institute 262
Andreev, Nikolai G. 231, 261
Andreev, Nikolai N. 262
Andrianov, Kuzma A. 50, 280
Andrusov, Nikolai I. 119, 324
Annex of the State Tretyakov Gallery → Tretyakov Filial

*There are no specific entries of the Academy of Sciences and of Lomonosov University (Moscow University) due to their ubiquity throughout the book. Page numbers followed by "n" refer to footnotes.

Anokhin Institute of Normal Physiology 183
Anokhin, Petr K. 182, 183, 291
Anosov, Pavel P. 142
anti-ballistic missiles 16, 151, 159
anti-matter 9
anti-science views/campaign 38, 63, 78, 93, 134, 240, 289
anti-Semitic campaign 14, 37, 44, 78, 184, 187, 349
anti-Semitism (anti-Jewish actions) 26, 87, 96, 203, 264, 284, 289
anti-Soviet activities (alleged) 10, 12, 16, 69, 208, 230, 232, 233, 281, 350
Anuchin Museum of Anthropology 97
Anuchin, Dmitry N. 97, 115, 125
Apollo-Soyuz project 174
Archimedes 78
Arctic Institute 301, 302
Arefiev, V. V. 31
Arkadiev, Vladimir K. 262
Arkhangelsky, Aleksandr A. 272
Armenian Academy of Sciences 320
Arnold, Vladimir I. 29, 258, 259
Arrhenius, Svante 242
Arseniev, Vladimir K. 127
artificial insemination 222
artificial organs 200, 201, 203, 205
artificial radioactivity 261
Artillery Academy 142
Artobolevsky, Ivan I. 46, 47, 277, 338
Artsimovich, Lev A. 20, 21, 24, 262
Arutyunov, Aleksandr I. 209, 210
Arzamas-16 8, 9, 17, 24, 28, 37, 90, 279, 280, 338, 348, 349
Astaurov, Boris L. 65, 286
Astronomical Society of the Pacific 320
Astrophysical Institute 261
atomic bomb project 8, 16, 18, 19, 24, 25, 31, 32, 36–38, 72, 91, 262, 263, 343
Auerbach, Charlotte 39
Authority of Mineral Resources 301
Authority of the Northern Sea Route 301
Averbakh Clinic of Ophthalmology 219
Averbakh, Mikhail I. 219
Averyanov, Sergei F. 243, 286
Avicenna 312

B

Babakin, Georgy N. 269
Bacteriological Institute (Odessa) 196
bacteriological weapons 199
Badalov, Ashot L. 336, 337
Baev, Aleksandr A. 63
Bagdasarov, Andrei A. 291
Baikov Institute 44, 45
Baikov, Aleksandr A. 18, 44 45, 142, 281
Bakh Institute of Biochemistry 54, 55, 63, 288
Bakh, Aleksei N. 53–55, 136, 280, 281
Bakulev Institute of Cardiovascular Surgery 335
Bakulev Research Center of Cardiovascular Surgery 281, 291
Bakulev, Aleksandr N. 221, 290, 291
Balandin, Aleksei A. 48, 280, 281
Baranov Institute of Aviation Motors 274
Baranov, S. A. 25
Baransky, Nikolai N. 298
Bardakh, Yakov Yu. 196
Bardin Institute of Iron Metallurgy 45
Bardin, Ivan P. 45, 142, 277
Bari, Nina K. 349
Barmin, Vladimir P. 152, 168, 269
Barnard, Christiaan 201
Barsukov, Valery L. 298
Basov, Nikolai G. 11–13, 262
Bauman University 6, 27, 45, 47, 50, 131, 132, 134, 145–157, 158, 159, 168, 170, 271, 277, 343
Bauman, Nikolai E. 146

Begichev, Nikifor A. 103
Belarusky University (Minsk) 20
Belorussian Academy of Sciences (BAS) 339
Belov, Nikolai V. 110
Belozersky Institute 95, 96
Belozersky, Andrei N. 95, 98, 286
Belyaev, Pavel I. 166
Belyakov, Rostislav A. 272, 273
Belyankin, Dmitry S. 298
Berg, Lev S. 122
Beriya, Lavrenty P. 17, 18, 31, 32, 277, 343
Berlin Wall 308
Berlin-Buch research center 64, 65
Bernshtein, S. N. 85
Berthelot, Marcellin 240
Bidloo, Nikolai (Nicolaas) 206
Blagonravov Institute of Machine Science 45, 47
Blagonravov, Anatoly A. 46, 47, 269
Blinova, Ekaterina N. 299
Blokhin Center of Oncology 334
Blokhin, Nikolai N. 291, 334, 335
Blokhintsev, Dmitry I. 21
Bochvar Institute for Inorganic Materials 150
Bochvar, Anatoly M. 142, 149, 150
Bochvar, Andrei A. 150, 338
Bogdanov, Anatoly P. 326
Bogdanov, Elly A. 239
Bogolyubov, Nikolai N. 88–90, 262
Bohr, Niels 34, 313, 314
Boky, Boris I. 61, 62
Bolshevik dictatorship 8
Bolshevik revolution 62, 238, 240
Boltinsky, Vasily N. 226
Bolyai, János 82
Bonaparte, Napoleon 236
Bonner, Elena 9
Boreisha, Maria 286
Borisyak Institute of Paleontology 118, 323, 325
Borisyak, Aleksei A. 118, 298, 324
Borovik-Romanov, Andrei S. 31, 262
Borovik-Romanov, Viktor-Andrei →Borovik-Romanov, Andrei S.
Borzov, Aleksandr A. 124, 298
Botanical Garden →Tsitsin Main Botanical Garden
Botanical Garden of Moscow University 96
Botanical Garden of St. Petersburg 124
Botanical Journal 340
Botkin Hospital 77, 208, 214–216
Botkin, Sergei P. 179, 214–216
Braunshtein, Aleksandr E. 63
Bredikhin, Fedor A. 91
Brezhnev, Leonid I. 20, 253
Britske, Edgar V. 281
Bruno, Giordano 177
Bryansk Institute of Transportation Machines 47
buckminsterfullerene 49
Budenny, Semen M. 237, 238
Budnikov, Petr P. 281
Budylkin, Gennady I. 233
Bukharin, Nikolai I. 54
Bunsen, Robert W. 240
Burakovsky, Vladimir I. 334, 335
Burdenko Clinic of Surgery 183, 207, 213
Burdenko Commission 207, 208
Burdenko Institute of Neurosurgery 207, 209, 210
Burdenko Military Hospital 205–208
Burdenko, Nikolai N. 199, 205–210, 291
Butlerov, Aleksandr M. 78, 92, 94
Butovich, Yakov I. 238

C

case of "doctors' plot" 208, 209, 221, 295–297
catastrophe theory 298
Catherine, Empress 70, 148

Cavendish Laboratory 30
Central Economics and Mathematics Institute → Central Institute of Economics-Mathematics
Central House of Scientists 341
Central House of the Artist 314
Central Institute of Economics-Mathematics 328, 339
Central Institute of Epidemiology and Microbiology 199
Central Post-Graduate Medical Institute 346
Central Telegraph 336, 337
Ceremonial Hall of Moscow University 75–77, 79
Chadwick, James 37
Chaplygin, Sergei A. 70, 146, 149
Charles University (Prague) 324
Chayanov, Aleksandr V. 232, 233
Chebyshev, Pafnuty L. 82, 83, 146, 149
Chekhov Arts Theater 180
Chekhov, Anton P. 179, 180, 191
Chelomei, Vladimir N. 156, 169, 170, 269, 276, 278
Chelyabinsk-70, 17, 279
chemical mutagenesis 38
Chemistry and Life 137
Cheremukhin, Aleksei M. 273
Cheremukhin, Georgy A. 273
Cherenkov Effect 7, 12, 263
Cherenkov, Pavel A. 7, 12, 263
Chernobyl 19, 23, 316, 346, 347
Chernogolovka 34, 40, 41
Chernov, Dmitry K. 142
Chernyshev, Fedosy N. 118
Chernyshevsky, Nikolai G. 80, 81
Chersky (Czerski), Ivan D. 127
Chertok, Boris E. 158, 269, 270
Chetvernikov, Sergei S. 65
Chichibabin, Aleksei E. 48, 50
Chizhevsky, Nikolai P. 281
Choibalsan, Khorloogiin 291
Chudakov, Evgeny A. 46, 152, 263
Chugaev, Lev A. 42, 48
Civil War 15, 19, 47, 62, 75, 91, 133, 154, 216, 217, 238, 271, 293, 294, 326
Cockcroft, John D. 319
collapse of the Soviet Union 86, 140, 251, 253, 259
College de France 144
combination scattering →Raman Effect
Communist Academy 171
Computational Center of the University 260
Copernicus, N. 78, 177
Copley Medal 77
Courant, Richard 86, 259
Course of Theoretical Physics (Landau and Lifshits) 34
Crafts School →Bauman University
Crimean University (Simferopol) 17, 17n
Crimean war 190, 214
Curie, Irène 7
Curie, Marie 12, 78
Cuvier, Georges 324

D

Dalton, John 78
Darwin Museum 108, 320–322, 326
Darwin, Charles 78, 104, 108, 119, 240, 321, 324
Davidenko, V. A. 25
Davydovsky, Ippolit V. 210, 211, 291
Day of Aviation and Cosmonautics 161
Demikhov, Vladimir P. 182, 199–202, 204, 348
Demiyanov, Nikolai Ya. 241, 242
Department of Normal Physiology 182, 183
Descartes, René 78
Deviche Pole campus 70, 178–180, 182, 183–199
Dezhnev Cape 122
Dezhnev, Semen I. 122

Dikushin, Vladimir I. 277
Dirac, Paul 7
Dobrovolsky, Georgy T. 342
Dobrynin, Vladimir A. 233
dogs in space 162–164, 173
Dokuchaev, Vasily V. 80, 81, 120, 121, 247
Dolgorukov family 311
Dollo, Louis 324
Donskoe Cemetery 193, 344
Dorodnitsyn Computing Center 27
Dorodnitsyn, Anatoly A. 27, 29, 87, 259
Doyarenko, Aleksei G. 230, 246
Drosophila melanogaster 38
Dubovitsky, Fedor I. 40, 41
Dyakov, Mikhail I. 286
Dyatkina, Mirra E. 43
Dygai, N. A. 332
Dzerzhinsky, Felix E. 305, 306

E
Earth Science Museum 84, 100–129, 321
Eastern Orthodox Church →Holy Dmitry Prilutsky Eastern Orthodox Church
Edelshtein, Vitaly I. 244, 286
Egorov, Boris B. 210, 291
Egorov, Boris G. 209, 210, 291
Ehrenfest, Paul 34
Einstein, Albert 148, 177, 312–314, 317
Elansky, Nikolai N. 192, 213, 291
Electrotechnical Institute (Leningrad) 311
Elizabeth, Empress 67, 68, 84, 102
Emanuel Institute of Biochemical Physics 35
Emanuel, Nikolai M. 35, 93, 282
Engels, Friedrich 267
Enikolopov, N. S. 39, 282
Erdgeschichte (Neumayr) 325
Erisman Research Center of Hygiene 192
Erisman, Fedor F. (Friedrich H.) 191, 192
Ermoleva, Zinaida V. 198, 199, 345, 346
Euler, Leonhard 10, 26, 78
Evdokimov Medical University of Stomatology 181, 217
Evdokimov, Aleksandr I. 217
exiles 4, 9, 10, 14, 32, 44, 54, 70, 71, 87, 120, 127, 139, 140, 144, 186, 203, 223, 228, 230, 232, 233, 264, 267, 272, 278, 281, 289, 305, 324, 333
Experimental transplantation of vital organs (Demikhov) 200

F
Faculty of Biology and Soil Science 63
Faculty of Biology 96–99, 122, 200, 288, 341
Faculty of Chemistry 22, 66, 92–96, 281
Faculty of Computational Mathematics and Cybernetics 86, 87, 260
Faculty of Geography 84
Faculty of Geology 84
Faculty of Mechanics and Mathematics 84–87
Faculty of Medicine (Moscow University) 62, 68, 192
Faculty of Physics 8, 14, 41, 66, 72, 87–92, 104, 266, 269
Faraday, Michael 78
Farnsworth, Philo T. 145n
Fedchenko, Aleksei P. 126
Federal Agency of Scientific Organizations (FASO) 2, 4
federal laws 2013 and 2014 3–5
Fedorov Institute of Microsurgery of the Eye 218
Fedorov, Efgraf F. 111

Fedorov, Evgeny K. 298
Fedorov, Ivan 330
Fedorov, Svyatoslav N. 182, 217, 218
Fedorov, Vladimir D. 292
Fedorova Maina V. 292
Fersman Mineralogical Museum 113, 322
Fersman, Aleksandr E. 109, 112, 113, 298, 312
Fesenkov, Vasily G. 264, 272
FIAN →Lebedev Institute
Filatov Municipal Clinical Hospital 193
Filatov Pediatric Hospital 193
Filatov, Nil F.178, 193
Filimonov, S. I. 31
First Moscow Medical Institute →Sechenov Medical University
First Moscow Pediatric Hospital →Morozov Pediatric Hospital
First Municipal Hospital 218, 220
Flerov, Gregory N. 16, 18, 25, 263
Flerovium (Fl) 263
Flexner, Simon 316
Florey, Howard 346
Fokin, Aleksandr V. 49, 50
Fortov, Vladimir E. 341
Fourth Medical Institute 181
Frank, Ilya M. 7, 12, 13, 88, 350
French Academy of Sciences 96
Fridman (Friedman, Friedmann), Aleksandr 34, 299
Frolov, Konstantin F. 46, 47
Frumkin Institute 43, 44
Frumkin, Aleksandr N. 43, 44, 136, 281, 282, 318
Frumkin, Anatoly P. 216

G

Gagarin Air Force Academy → Zhukovsky-Gagarin Air Force Academy
Gagarin, Yury A. 148, 160–162, 165, 166, 173, 342
Galilei, Galileo 78, 148, 177
Galpern, Elena G. 49
Gamaleya Center of Epidemiology and Microbiology 197, 198
Gamaleya, Nikolai F. 188, 196, 197, 292
Gannushkin, Petr B. 292
Gapon, Evgeny N. 241, 242
Gaudi, Antoni 134
Gauss, Carl Friedrich 78
Gazenko, Oleg G. 163, 164
Gedroits, Konstantin K. 120, 123
Gelfand, Izrail M. 95, 96
genetic code 79
Geographical Society 57, 107, 117, 198
Gerie (Guerrier), Vladimir I. 347, 348
German aggression 16, 18, 38, 40, 199
Gertsen Institute of Oncology 196, 292
Gertsen Oncological Institute → Gertsen Institute of Oncology
Gertsen, Petr A. 210, 292
Getye, Fedor A. 215, 216
Gilbert, Walter 64
Gilyarov, Merkury S. 56, 287
Gindtse Museum of Anatomy 236
Gindtse, Boris K. 237
Ginzburg, S. Z. 332
Ginzburg, Vitaly L. 12, 14, 15, 24, 88, 264, 267, 268, 318
Gladkov, Georgy A. 24
Glagoleva-Arkadieva, Aleksandra A. 262
Glavmosstroi 276
Glushko, Valentin P. 168, 169, 170, 172, 272
GOELRO Plan 112
Gogol, Nikolai 255
Goldansky, Vitaly I. 39, 282, 286
Golitsin, Boris B. 106, 107
Goncharov, V. V. 25
Gorbachev, Mikhail S. 10, 23, 208, 316
Gorkov, A. P. 31
Gorkovsky Medical Institute → Lobachevsky University
Goryachkin Agro-engineering University 226

Goryachkin, Vasily P. 227
Gosplan 117
Grigoryan, Samvel S. 320
Groves, Leslie 280
Grozny Petroleum Institute 21
Grushin, Petr D. 270
Gubkin, Ivan M. 114, 137–139, 298
Gubkin University/Institute 137, 138, 285, 300
Gudtsov, Nikolai T. 282
GULAG 73, 306
Gulevich, Vladimir S. 68, 282
GUM 132
Gusev, Leonid I. 172
Gusev, Mikhail V. 98
Gustavson, Gavriil G. 241, 242

H
Harvard University 64
Helmholtz Institute 219, 219n
Helmholtz, Hermann von 187, 240
Herzen, Aleksandr A. 292
Herzen, Aleksandr I. 70, 82, 292
Higher Courses for Women 65, 68, 97, 122, 149, 180, 181, 187, 219, 220, 240, 242, 285, 286, 295, 299, 321, 347, 348
Higher Technical School →Bauman University
Hilbert, David 86, 259, 260, 261
Hitler, Adolf 337
Hodgkin, Dorothy 42
Holy Dmitry Prilutsky Eastern Orthodox Church 183, 201, 202
House of the Government →House on the Embankment
House on the Embankment 205, 332, 333, 335
House on the Embankment (Trifonov) 335
Hubutiya, Mogeli Sh. 204
human rights activities 4, 9, 10, 14, 188, 305
Humboldt University 105, 241
Humboldt, Alexander von 105, 106
hybridization of humans and apes 222, 223
hydrogen bomb 8, 9, 10, 14, 18, 305

I
Ig Nobel Prize 51
IKAN →Shubnikov Institute
Ilichev, Aleksandr S. 298
Ilyin, Ivan A. 71
Ilyushin, Sergei V. 148, 273, 274
Imperial Clinical Institute (St. Petersburg) 185
Imperial Moscow Technical School → Bauman University
INEOS →Nesmeyanov Institute
Institut Pasteur 196, 198, 222, 223
Institute for Nuclear Research (Dubna) 263
Institute of Atomic Energy →Kurchatov Institute
Institute of Biomedical Problems 163, 164
Institute of Bioorganic Chemistry →Shemyakin-Ovchinnikov Institute
Institute of Biophysics 21, 289
Institute of Cosmic Biology and Medicine →Institute of Biomedical Problems
Institute of Crystallography (IKAN) →Shubnikov Institute
Institute of Cytology, Histology, and Embriology 38, 98
Institute of Electrical Engineering (Tbilisi) 337
Institute of Elemet-Organic Compounds (INEOS) →Nesmeyanov Institute
Institute of Energy Problems in Chemical Physics 40
Institute of Epidemiology, Microbiology, and Infectious Diseases 199

Institute of Exact Mechanics and Computational Technology 27
Institute of Experimental and Clinical Surgery 200
Institute of Experimental Biology → Koltsov Institute of Developmental Biology
Institute of Experimental Medicine 63, 77, 267
Institute of General Genetics →Nikolai Vavilov Institute
Institute of General Physics →Prokhorov Institute
Institute of Genetics 38, 54, 57, 252
Institute of Geochemistry 301
Institute of Geosciences 301
Institute of Hydrogen Energetics and Plasma Technologies 24
Institute of Hygiene →Erisman Research Center
Institute of Hygiene and Protection of the Health of Children and Adolescents 211
Institute of Internal Medicine →Myasnikov Institute of Cardiology
Institute of Mathematics (Novosibirsk) 261
Institute of Mathematics (Ukraine) 27
Institute of Mathematics and Mechanics 260, 261
Institute of Microbiology 198
Institute of Molecular Biology →Engelhardt Institute
Institute of Morphology 184
Institute of Natural Products Chemistry →Shemyakin-Ovchinnikov Institute
Institute of Oceanography 302
Institute of Oil →Gubkin University
Institute of Physical Chemical Analysis 42
Institute of Physical Problems →Kapitsa Institute
Institute of Physical Technology (Leningrad) →Ioffe Institute
Institute of Physics and Mathematics 12
Institute of Plant Breeding 57
Institute of Problems of Mechanics 299
Institute of Radiation and Physical Chemical Biology →Engelhardt Institute
Institute of Radio Technique 15
Institute of Scientific and Technical Information 50
Institute of Steel →University of Science and Technology
Institute of the History of Science and Technology 7
Institute of Theoretical Physics → Landau Institute
Institute of Virology 199
Institute of Zoology (Leningrad) 197
International Congress of Biochemistry 79
International Mathematical Union 86
Ioffe Institute of Physical Technology 16, 18–20, 23, 24, 34–36, 311
Ioffe, Abram I. 16–18, 23, 30, 34–36, 91
Ipatiev, Vladimir N. 48
Isaev, Aleksei M. 151, 270
Ishlinsky, Aleksandr Yu. 319, 320
Ivanenko, Dmitry D. 88, 242
Ivanov, Ilya I. 222, 223
Ivanov, Leonid A. 287
Ivanov, Mikhail F. 236

J
Jewish Anti-Fascist (Antifascist) Committee 44, 289
Johnson, Lyndon B. 145
Joliot-Curie, Frédéric 7

K
Kabachnik, Martin I. 50
Kabanov, Viktor A. 95, 282

Kablukov, Ivan A. 241, 242
Kadomtsev, Boris B. 21
Kaftanov, Sergei V. 134, 282
Kalinin, Mikhail I. 248
Kamensky, Grigory N. 299
Kamov, Nikolai I. 274
Kantorovich, Leonid 29, 261
Kapitsa Institute 19, 29, 30, 32–34, 36, 262, 291
Kapitsa Museum 29, 32, 320
Kapitsa, Andrei P. 31, 33
Kapitsa, Anna A. 31, 32, 286
Kapitsa, Petr L. 19, 20, 29–34, 69, 85, 88, 91, 262, 264–266, 286, 315, 319
Kapitsa, Sergei P. 31, 33
Kapustinsky, Anatoly F. 282
Kargin, Valentin A. 94, 95, 136, 283, 285
Kármán, Theodore von 149
Karpinsky, Aleksandr P. 62, 106, 112, 342
Karpov Institute of Physical Chemistry 44, 136, 137
Karpov, Lev Ya. 136
Kassirsky, Iosif A. 293
Kasso, Lev A. 6, 70
Katkov, Georgy F. 270
Katyn massacre 208
Katyusha multiple rocket launchers 28, 152, 286
Kazan University 24, 63, 82, 92, 94, 186, 294
Kazansky, Boris A. 283
Kedrov, Bonifaty M. 283
Keldysh Institute of Applied Mathematics 26, 27, 37, 96, 170, 350
Keldysh, Mstyslav V. 26, 29, 75, 149, 169, 170, 272, 342, 350
Kepler, Johannes 78, 177
KGB 10, 14, 87, 306
Khaikin, M. S. 31
Khalatnikov, I. M. 31
Khariton, Mariya N. 265
Khariton, Tatyana Yu. 265, 286
Khariton, Yuly B. 22, 24, 28, 37, 265, 286, 338, 343
Kharkevich Institute for Information Transmission Problems 265
Kharkevich, Aleksandr A. 265
Kharkov University 98, 222
Khinchin, A. Ya. 85
Khitrin, Lev N. 266
Khlebutyn, Evgeny B. 233
Khlopin Radium Institute 112
Khokhlov, Rem V. 90, 91, 266
Khoroshko, Vasily K. 293
Khristianovich, Sergei A. 27, 28, 91, 149, 261, 348
Khrushchev, Leonid N. 256
Khrushchev, Nikita S. 5, 39, 51, 87, 95, 253, 256, 257
Khudyakov, Nikolai N. 57, 238, 239
Kibel, Ilya A. 299
Kiev Aviation Institute 170
Kiev Polytechnic Institute 11, 45, 168, 276, 281, 283
Kiev University 19, 54, 56, 90, 97, 301
Kikoin, Isaak K. 23, 24, 266, 343
Kirchhoff, Gustav 240
Kirillin, Vladimir A. 158, 159, 277
Kiselev, Fedor L. 63, 199, 290
Kiselev, Lev L. 62, 63, 199, 290
Kiseleva, Valeriya P. 198
Kistyakovsky, Vladimir A. 43, 44, 283
Kitaigorodsky, Aleksandr I. 49, 50, 283, 284
Kitaigorodsky, Isaak I. 134, 283, 284
Klechkovsky, Vsevolod M. 245, 246, 348
Kleimenov, I. T. 272
Klein, Felix 260
Klimov, Vladimir Ya. 274
Klyucheva, Nina 186, 187, 297, 335
Knipovich, Nikolai M. 116, 117
Knunyants, Ivan L. 50

Kochergin, I. G. 218, 219
Kochin, Nikolai E. 29, 149, 299
Kochina, Pelageya Ya. 299
Koksharov, Nikolai I. 114
Kolesnev, Samuil G. 233
Kolesnikov, Benedikt A. 244
Kolmogorov, Andrei N. 29, 85, 86, 258, 259
Kolotyrkin, Yakov M. 137
Koltsov Institute of Developmental Biology 38, 64, 65, 288, 334
Koltsov, Nikolai K. 38, 62, 64, 65, 350
Komarov, Vladimir L. 54, 121, 272, 287
Komarov, Vladimir M. 162, 166, 342
Kondratiev, Nikolai D. 232, 233
Kondratiev, Viktor N. 36
Konenkov Studio-Museum 316, 317
Konovalov, Dmitry 45
Konovalov, Nikolai V. 293
Konstantinov, Petr N. 231
Korobov, Pavel I. 284
Korolenko, Vladimir G. 248
Korolev Museum 169
Korolev, Sergei P. 152, 153, 157, 167–169, 172, 175, 271, 272, 343
Korsakov Clinic of Nervous Diseases 195
Korsakov, Sergei S. 195
Korshak, Vasily V. 50
Kosberg, Semen A. 274
Kost, Ekaterina A. 295
Kostandov, Leonid A. 135, 136, 343
Kostyakov Institute of Soil Melioration and Hydro-construction 243
Kostyakov, Aleksei N. 243
Kostychev, Pavel A. 121
Kotelnikov Institute of Radio Physics and Electronics 280
Kotelnikov, Gleb E. 273, 274
Kotelnikova, Ada S. 43
Kots, Aleksandr F. (Alexander E. Kohts) 321, 326
Kovalenko Institute of Experimental Veterinary Science 222, 223
Kovalenko, Yakov R. 222, 223
Kovalevskaya, Sofiya V. 118, 119
Kovalevsky, Vladimir O. 118, 119
Kozhevnikov, A. Ya. 195
Kraevsky, Aleksandr A. 63
Krasheninnikov, Ippolit M. 124
Krasheninnikov, Stepan P. 106
Krasnobaev, Timofei P. 194
Krasnov, Andrei N. 124, 128
Kremlin Hospital 208, 216, 221, 296
Kremlin Wall 136, 161, 162, 341–343
Krenkel, Ernst T. 298, 299, 300
Krishtofovich, Afrikan N. 324
Kropotkin, Petr A. 119, 120, 299
Kruber, Aleksandr A. 115
Krug, Karl A. 158, 159
Krylov Institute 140
Krylov, Aleksandr P. 140, 141
Krylov, Aleksei N. 29, 33, 85, 315
Krylov, Nikolai M. 259
Krzhizhanovsky Energy Institute 266
Kulagin Zoological Museum 236
Kulagin, Nikolai M. 97, 98, 237
Kulebakin, Viktor S. 274
Kuleshov, Pavel N. 240
Kuntsevskoe Cemetery 266, 345
Kurchatov Institute 8, 16, 17, 19–22, 24, 25
Kurchatov, Boris V. 24, 284
Kurchatov, Igor V. 16–19, 91, 263, 343
Kurnakov Institute 42, 43, 44, 286
Kurnakov, Nikolai S. 42, 85, 281
Kursanov, Andrei L. 287
Kursanov, Lev I. 287
Kursanov, Nikolai I. 287
Kuzin, Mikhail I. 212, 213
Kuzminskoe Cemetery 345, 346
Kuzminsky Park 222, 327
Kuzminsky-Lyublino Park → Kuzminsky Park
Kuznetsov, Boris A. 237
Kuznetsov, Viktor I. 168, 170, 171, 270

L

Laboratory No. 2 →Kurchatov Institute
Laboratory of Experimental Zoology 98
Laboratory of Measuring Instruments → Kurchatov Institute
Laboratory of Medicinal Chemistry 68
Lake Ladoga 299
Lamarck, Jean-Baptist 59, 321
Lamb, Willis E. 319
Landau Institute 34
Landau, Edmund 261
Landau, Lev D. 14, 30–34, 88, 266, 291, 307
Landsberg, Grigory S. 88, 266, 267
Lange, Oktavy K. 116
Langemak, G. E. 272
Langevin, Paul 144
Langman, A. Ya. 332
Lausanne medical school 333
Lavochkin, Semen A. 270, 271
Lavoisier, Antoine 111
Lavrentiev, Mikhail A. 27–29, 149, 170, 260
Lazarev, Petr P. 15, 266, 267
Le Corbusier 134
Lebedev Institute (FIAN) 5–16, 21, 24, 26, 262, 269
Lebedev Institute of Precise Mechanics and Computer Engineering 159
Lebedev, Petr N. 5, 6, 10, 45, 69, 70, 88–90, 146, 262, 266, 267
Lebedev, Sergei A. 159, 259
Lebedev, Valentin V. 166
Lebedinsky, Andrei V. 163, 293
Lefortovo prison 203
Legasov, Valery A. 22, 23, 134, 284
Leggett, Anthony J. 14
Leibenzon, Leonid S. 137, 149, 277, 278
Leichoudes brothers 330, 331
Leiden University 105
Lenin Academy of Agricultural Sciences →Academy of Agricultural Sciences
Lenin Mausoleum 297, 342
Lenin, Vladimir I. 5, 231, 297
Leningrad Polytechnic Institute → Polytechnic Institute
Leningrad University 13, 28, 34, 38, 63, 170, 260, 303, 320, 340
Leonov, Aleksei A. 166, 167
Leontovich, Aleksandr V. 293
Leontovich, Mikhail A. 11, 13, 20, 21, 88
Lepeshinskaya, Olga B. 55, 69, 333, 334
Lepeshinsky, Panteleimon N. 333
letters of protest and condemnation 14, 20, 31, 54, 85, 87, 311
Levene, Phoebus A. 316
Levi-Civita, Tullio 149
Levinson-Lessing, Frants Yu. 106
Library of Moscow University 83, 84
Lifshits, Evgeny M. 31, 34, 266
Lilienthal, Otto 147
Linné, Karl 57, 105
Lisitsyn, Petr I. 227, 228, 230
Liskun, Efim F. 240, 287
Litke, Fedor P. 116, 117
Livanov, Dmitry V. 143
Lobachevsky University (Nizhny Novgorod) 334
Lobachevsky, Nikolai I. 10, 78, 80, 81, 82
Lobashov, Mikhail 340
Lomonosov Institute of Fine Chemical Technologies 63, 137, 285
Lomonosov Moscow Auto-Tractor Institute 46
Lomonosov, Mikhail V. 10, 42, 66–68, 71, 74, 78, 82, 83, 103, 106, 111, 129, 137, 142, 177, 313
Los Alamos 18
Loza, Grigory M. 233
Lubyanka 33, 203, 306, 307, 316

Luzhniki sports arena 80, 278
Luzin case 85
Luzin, Nikolai N. 27, 29, 85, 170, 259, 349, 350
Lyapunov, Aleksandr M. 350
Lyapunov, Aleksei A. 349, 350
Lysenko, Trofim D. and his impact 38, 51, 55, 57–60, 63, 69, 79, 87, 134, 139, 228, 240, 252, 253, 285, 288, 290, 325, 334, 339, 340, 345
Lyubimova, Anna K. 40
Lyulka, Arkhip M. 274

M
Maisuryan, Nikolai A. 230
Makarov, Stepan O. 117
Makeev, Viktor P. 271
Malkov, M. P. 31
Malyshev, Vyacheslav A. 343
Mandelshtam, Leonid I. 8, 14, 21, 30, 88, 266, 267
Manhattan Project 280
Marchuk, Gury I. 259
Marconi, Guglielmo 81
Mardashev, Sergei R. 293
Maritime Ministry 302
Markov, Moisei A. 11
Markovnikov, Vladimir V. 93, 94, 245
Marxism-Leninism 87, 93, 261, 313
Master and Margarita (Bulgakov) 208
Mathematical Cabinet of the Academy of Sciences 26
Mathematical Institute (MIAN) → Steklov Institute
mathematical linguistics 86
Matrosov, Ivan K. 278
Matthaei, Heinrich 79
Maxwell, James C. 6, 78
Mayo Clinic 201
Mazarovich, Aleksandr N. 118, 300
Mechnikov, Ilya (Élie) I. 196
Medgamal 197, 198
Medical University of Stomatology → Evdokimov Medical University of Stomatology
Medvedev, Dmitry A. 249
Medvedev, Sergei S. 136, 137, 285
Melnikov, Nikolai P. 278
Memorial Museum of Cosmonautics 163, 173–177
Mendel, Gregor 60, 240
Mendeleev University 22, 50, 134, 135, 149, 281, 328
Mendeleev University of Chemical Technology →Mendeleev University
Mendeleev, Dmitry I. 10, 42, 45, 57, 76, 77, 82, 92, 112, 134, 135, 137, 146, 149, 171, 187, 240, 250, 328
Mendelevium (Md) 77
Menzbir, Mikhail A. 64, 97, 325
Merkin, Vladimir I. 25
Mersalov, Nikolai I. 151
meteorology 126, 228, 229, 299
metro stations 328, 329
Michurin, Ivan V. 58, 59, 76, 77, 80, 81, 123
Middendorf, Aleksandr F. 127
Mikdal, Arkady B. 268
Mikheev, Mikhail A. 278
Mikhelson Meteorological Observatory 228
Mikhelson, Vladimir A. 228, 229
Mikoyan, Anastas, I. 335
Mikoyan, Artem I. 148, 274, 335
Milanovsky, Evgeny V. 115, 300
Military Academy of Chemical Defense 50
Military Academy of Radiation, Chemical, and Biological Defense 50
Military Medical Academy 163, 196, 197, 209, 213, 335
Military Medical Service 209, 216
Millionshchikov, Mikhail D. 21, 278
Minakov, N. N. 31

Mineralogical Society 111
Mining Academy (Moscow) 45, 107, 116, 118, 122, 137, 138, 282, 343
Mining Institute (Moscow) 51, 61, 143, 300, 301, 303
Mining Institute (St. Petersburg/ Leningrad) 42, 61, 62, 107, 111, 118, 138, 141, 259, 302, 303
Ministry of Aviation Industry 270
Ministry of Education 5, 281, 340, 341
Ministry of Foreign Affairs 73
Ministry of General Machine Building 172
Ministry of Health 199, 213, 267, 291
Ministry of Medium Machine Building 280, 343
Ministry of Science and Higher Education 340, 341
Ministry of the Interior and State Security 188, 343
Minsk Polytechnic Institute 311
Mints Institute of Radio Technique 15
Mints, Aleksandr L. 15, 278
Mirchink, Georgy F. 119, 120
Mironov, Stepan I. 300
Mirzabekov, Andrei D. 62–64, 348
Mirzo Ulugbek University 62
MISiS →University of Science and Technology
Mitinskoe Cemetery 20, 346, 347
Mobius bands 327, 328, 339
Molchanov, Vasily I. 293
Molecular Biology 63
molecular biology 53, 61, 64, 79, 98
Molotov, Vyacheslav M. 85
Morgan, Thomas Hunt 60, 290
Morozov Pediatric Hospital 194
Morozov, Georgy F. 122
Moscow Agricultural Institute → Timiryazev Academy
Moscow Aviation Institute 47
Moscow City Hospital →Third Medical Institute
Moscow Clinical Pediatric Hospital 295
Moscow Energy Institute 11, 12, 24, 158, 159, 172
Moscow Geological School 125
Moscow Institute of Chemical Machine Construction 25
Moscow Institute of International Relations 92n
Moscow Institute of Physical Engineering 11, 12, 21, 25, 268
Moscow Institute of Physics and Technology (*PhysTech*) 14, 21, 24, 27, 28, 34, 40, 91, 93, 205, 266, 279
Moscow Mathematical Society 86
Moscow Medinstitute 181
Moscow Pedagogical University 56, 180
Moscow Regional Hospital →Fourth Medical Institute
Moscow Society of Naturalists 90, 97
Moscow University tower 2, 73–75, 80, 83, 84, 102
Muromtsev, Georgy S. 287, 288
Muromtsev, Sergei N. 287, 288
Museum of Animal Husbandry 236
Museum of Bauman University 156, 157
Museum of Horse-breeding 236–238
Museum of Modern Art (Moscow) 309
Museum of Psychiatry 188, 321
Museum of the Faculty of Physics 88
Museum of the History of Medicine 133, 187–189, 191, 321
Muzeon Park 304, 305, 306, 311–315
Myasnikov Institute of Cardiology 214
Myasnikov, Aleksandr L. 179, 213, 214, 293

N

Nadiradze, Aleksandr D. 271
Nametkin, Sergei 48, 137, 285
Nansen, Fridtjof 331, 332
Nasonov, Nikolai V. 98
National Air and Space Museum (Washington, DC) 174

National Broadcasting Company (NBC) 145
National Institutes of Health (NIH, US) 79
National Research Center "Kurchatovsky Institute" → Kurchatov Institute
nature preserves 69
Nazarov, Ivan N. 47
Nazis 71, 188, 337
Negul, Aleksandr M. 244
Nemchinov, Vasily S. 252, 338, 339
Nemytsky, Viktor V. 349
Nesmeyanov Institute (INEOS) 48–51
Nesmeyanov, Aleksandr N. 5, 48, 49, 51, 53, 69, 73, 93, 101, 284
Neumayr, Melchior 324, 325
Neustruev, Sergei S. 123
Nevelskoi, Gennady I. 128
New York Scientific (Hargittai and Hargittai) 316
Newton, Isaac 78, 129, 148, 177
Niemeyer, Oscar 134
Nikitin, Nikolai V. 278
Nikitin, Sergei N. 119
Nikitin, Vasily P. 278
Nikogosyan Art Gallery 319
Nikolaev, Georgy A. 151
Nikolai I. Vavilov Society 65
Nikolai Vavilov Institute 57
Nikolai Vavilov Museum (at Timiryazev Academy) 228
Nirenberg, Marshall W. 79
Nobel Committees 39
Nobel Foundation 39
Nobel Peace Prize 9, 88, 331
Nobel Prizes 7, 7n, 11–15, 24, 30, 34, 37, 39, 49, 64, 76, 77, 88, 143, 184, 196, 267, 311, 317, 319, 341
Nobile, Umberto 149
Noguchi, Hideyo 316
North Pole expedition 298, 299, 300
Novodeviche Monastery 254, 255
Novorossiisky University (Odessa) 44, 48, 90, 94, 98, 187, 196, 324
nuclear explosion (1954) 213
nuclear power plants 18, 19, 23, 25, 347
nuclear projects/weapons 7, 10, 16–22, 24, 25, 27, 28, 34, 37, 40, 65, 84, 137, 150, 171, 172, 213, 246, 260, 261, 263, 268, 269, 279, 280, 303, 343, 346, 348 (see also thermonuclear)

O

Oberli, Leon 199
Obruchev, Vladimir A. 106–108, 116, 300
Obrucheva, Amaliya D. 282
Obukhov, V. S. 25
Ochkin, Aleksei D. 216, 292
Ogarev, Nikolai P. 70
oil exploration 139, 140
Okolesnov, S. P. 31
Oparin, Aleksandr I. 54, 55, 69, 288, 334
Orekhovich Institute of Biomedicinal Chemistry 189
Orekhovich, Vasily N. 188, 189
origin of life 55
Orlov Museum of Paleontology 323
Orlov, Yury A. 288, 324, 325
Ostankino TV tower 278
Ostroumov Clinic of Internal Medicine 183
Ostwald, Wilhelm 242
Our Earth in the Universe (exposition) 104
Ovchinnikov, Yury A. 52, 53, 285
oxygen production 31, 143, 265

P

Paleontological Society 325
Pallas, Peter-Simon 105
Panasyuk, I. S. 25

Papaleksi, Nikolai D. 13
Papanin, Ivan D. 298–300, 302
Parasitological Institute 54
Parenago, Pavel P. 268
Parin, Vasily V. 163, 187, 293
Pasternak, Boris 7n, 39
Pasteur Institute →Institut Pasteur
Pasteur, Louis 78, 196, 250
Patsaev, Viktor I. 342
Pauling, Linus 43, 93
Pavlov Institute of Physiology 63
Pavlov Medical University (Ryazan) 181
Pavlov, A. I. 31
Pavlov, Aleksei P. 108, 109, 125, 325
Pavlov, Ivan P. 76–78, 80, 81, 183, 223, 317, 322
Pavlov, Mikhail A. 142
Pavlova, Mariya V. 109, 300, 324, 325
Pavlovsky, Evgeny N. 197
Peive, Aleksandr V. 300
penicillin 199, 346
Perevalova, Emiliya G. 282
Perevozchikov, V. I. 31
periodic table of the elements 76, 328
Perov, Sergei S. 285
Persianinov, Leonid S. 293, 294
Pervukhin, M. G. 343
Peter, the Great (Peter I) 4, 5, 205, 206, 225, 314
Petrograd University →Leningrad University
Petroleum Institute (Grozny) 21
Petroleum Institute (Moscow) 282
Petrov, Boris A. 203
Petrov, Boris N. 271
Petrovsky Academy →Timiryazev Academy
Petrovsky Palace 148
Petrovsky, Boris V. 184, 205, 293
Petrovsky, Ivan G. 75, 259
Petryanov-Sokolov, Igor V. 137
Petrzhak, Konstantin A. 18
Philosophical Steamship 71
Physical Institute (FIAN) →Lebedev Institute
Physics of the Microworld (Shchelkin) 279
Picasso's style 43
Pilnyak, Boris A. 310
Pilyugin, Nikolai A. 152, 168, 170, 271
Pirogov Medical University 179, 181
Pirogov, Nikolai I. 179, 189, 190, 206, 214
Pitaevsky, L. P. 31
Planetarium 327
Plate, Nikolai A. 95, 285
Polikarpov, Nikolai N. 275
polymer science 19, 39, 50, 63, 94, 95, 137, 285, 296
Polytechnic Institute (St. Petersburg/Petrograd/Leningrad) 23, 30, 36, 42, 45, 47, 91, 144, 262, 274, 275
Polytechnical Museum 41, 89, 98, 160, 321
Pontecorvo, Bruno 337
Pontryagin, Lev S. 86, 87, 260
Popkov, Valery I. 278
Popov, Aleksandr S. 80, 81
Popov, Ivan S. 240
Popov, Petr P. 294
popular science 33, 56, 279
Pozdyunin, Valentin L. 279
Pradyuk, N. F. 31
Prandtl, Ludwig 149
Pravda 291
Preobrazhensky, Boris S. 220, 221, 294
presidents of the Academy of Sciences 4, 5, 7, 19, 21, 32, 51, 54, 62, 67, 73, 117, 170, 259, 341, 342
Prezent, Isai I. 59
Priorov, Nikolai N. 294
Privalov, Ivan I. 260
probability theory 83, 86, 259
Prokhorov Institute 13, 14
Prokhorov, Aleksandr M. 11–14, 87, 88, 268
Prokofiev, Mikhail A. 95
Prokofiev, Yu. A. 31

Prokofieva-Belgovskaya, Aleksandra A. 62, 63
Protodiyakonov, Mikhail M. 62
Prozorovsky, Sergei V. 197, 198
Pryanishnikov, Dmitry N. 57, 121, 230, 245, 246, 250
Przhevalsky, Nikolai M. 126
psychiatry in forced treatment 188
Pulkovo Observatory 91, 320
Puponin, Anatoly I. 231
Pushkin, Aleksandr S. 68
Putin, Vladimir V. 249
Pyatnitskoe Cemetery 347, 348

R
Raevskaya, Galina A. 296
Rakhmanov, Viktor A. 185, 186
Raman Effect 267
Raman, Chandrasekhara Venkata 267
Rapoport, Iosif A. 38, 39, 334, 340
Raspletin, Aleksandr A. 279
Razenkov, Ivan P. 182, 183
RCA (Radio Corporation of America) 145
Red Army →Soviet Army
Research Institute of Nuclear Physics of Moscow University 12, 269
Rockefeller University 316
Rockets and people (Chertok) 270
Rodionov, Vladimir M. 285
Roentgen Institute 30
Roentgen, W. C. 36
Rona, Peter 63
Roskin, Grigory 186, 187, 205, 297, 335
Royal Astronomical Society (UK) 320
Royal College of Surgeons (UK) 203
Royal Society (London) 77, 86, 96, 107
Royal Swedish Academy of Sciences 67, 76, 96
Royal Swedish Medical Society 216
Royal Uppsala Scientific Society 201
Rozing, Boris L. 144
Rubinin, P. E. 31
Rulie, Karl F. 102, 326
Runkle, John D. 149
Rusanov, Vladimir D. 24
Russian Academy of Arts 84
Russian House (Washington, DC) 309
Russian program "Human Genome" 63
Russian Red Cross 7, 189, 331
Russian Space System (corporation) 172
Russo-Japanese war 117, 207, 238
Rutgers University 96
Rutherford, Ernest 6, 30, 78
Ryazansky, Mikhail S. 168, 172
Rychkov, Petr I. 125
Rzhevsky, Vladimir V. 300

S
Sadovnikova-Koltsova, Mariya P. 350
Sakharov Archive 307, 309
Sakharov Center 308, 309
Sakharov Garden 308
Sakharov, Andrei D. 4, 7–10, 12, 14, 20, 87, 88, 158, 304–307, 310, 348
Sakharov, Boris A. 285
Salmanov, Farman K. 139, 140
Samarin, Aleksandr M. 44, 45, 285
Samoilovich, Rudolf L. 302
Sarkisov, Donat S. 212
Sarnoff, David 145
Satel, Eduard A. 154
Saukov, Aleksandr A. 300
Savitskaya, Svetlana E. 166
Sazhin, Nikolai P. 285
Schmidt, Otto Yu. 254, 300, 301
Science 339
Science Museum (London) 134
Scud rockets 151, 152
Sechenov Medical University 179, 182, 195, 196, 207, 296, 298
Sechenov, Ivan M. 77, 78, 97, 171, 179, 180, 182, 183, 187, 188, 214, 294
Second Moscow Medical Institute → Pirogov Medical University

Second Moscow University 181
Sedov, Leonid 149
Semashko, Nikolai A. 186, 187, 294
Semenov Institute of Chemical Physics 2, 35–41, 93, 338
Semenov, Aleksei Yu. 265
Semenov, Nikolai N. 35–41, 91, 93–95, 265, 285, 286
Semenov, Yury N. 265
Semenova, Natalya N. 265
Semenov-Tyan-Shansky, Petr P. 126
Semipalatinsk proving ground 18
Senkevich, Yury A. 294
Sepp, Evgeny K. 294
Serbinenko, Fedor A. 209, 210
Serebrovsky, Aleksandr S. 288
Sergeev, Sergei S. 233
Sergiev, Petr G. 294
Serzhantov, Aleksandr V. 326, 327
Severgin, Vasily M. 111
Severin, Sergei E. 99
Severtsov Institute of Ecology and Evolution 56, 99
Severtsov, Aleksei N. 53, 97
Severtsov, Nikolai A. 122
Shabarova, Zoya A. 95
Shabolovka Tower →Shukhov Tower
Shalnikov, Aleksandr I. 31, 36
Shanyavsky University 6, 15, 21, 41, 65
Shaposhnikov, Vladimir N. 288
sharashkas 155, 199, 272, 273
Shaternikov, Mikhail N. 182, 183
Shatilov, Ivan S. 231
Shchelkin, Kirill I. 279
Shcherbakov, Dmitry I. 301
Shchurovsky, Grigory E. 113, 114
Shemyakin, Mikhail M. 51–53, 285
Shemyakin-Ovchinnikov Institute 52, 53
Sheremetevo, Nikolai 202
Shevyakov, Lev D. 301
Shipilov, Vasily S. 236
Shirshov, Petr P. 298, 301, 302, 335
Shitt, Petr G. 244
Shkhvatsabaya, Igor K. 295
Shkhvatsabaya, Konstantin Ya. 295
Shkhvatsabaya, Tatiyana V. 295
Shkhvatsabaya, Yury K. 295
Shnoll, Simon E. 340
Shokalsky, Yuly M. 117
Shostakovich, Dmitry D. 223, 310
Shreder, Rikhard I. 226, 235
Shternberg Astronomical Institute 91
Shternberg, Pavel K. 91
Shubnikov Institute 41, 42
Shubnikov, Aleksei V. 41
Shukhov Tower 2, 130, 131, 133, 134
Shukhov, Vladimir G. 131–133, 146, 157, 278, 279
Shumakov Institute of Transplantology and Artificial Organs 200, 201, 204
Shumakov, Valery I. 204, 205, 295
Shuvalov, Ivan I. 67, 84, 102
Siberian Branch of the Academy of Sciences 27, 28, 261, 350
Sibirtsev, Nikolai M. 121
Sidorenko, Aleksandr V. 301, 302
Simonenko, D. L. 25
Sinyukov, Mikhail I. 233
Sisakyan, Norair M. 288
Sklifosovsky Institute of Emergency Services 200, 202–204
Sklifosovsky, Nikolai V. 179, 185
Sklodowska-Curie, Marie →Curie, Marie
Skobeltsyn, Dmitry V. 11, 12, 88, 268
Skryabin, Georgy K. 87, 289
Skryabin, Konstantin I. 54, 222, 223, 288, 289, 339
Skrzhinsky, Nikolai K. 276
Skvortsov, Mikhail A. 295
Slavyanov, Nikolai N. 302
Smirnov, Efim I. 208, 209, 295
Smirnov, Sergei S. 113
Snegirev Clinic of Obstetrics and Gynecology 192

Snegirev, Vladimir F. 192, 193
Sobolev, Sergei L. 260, 261
Society for the Acclimatization of Animals and Plants 98
Society of Surgeons (Paris) 203
Sofiya Pediatric Hospital →Filatov Pediatric Hospital
Sokolov, K. M. 332
Sokolov, Vladimir E. 56, 99
Soldatenkov, Kozma T. 215
Soldatenkovsky Hospital →Botkin Hospital
Soloviev, Vladimir A. 166
Sorbonne 112, 203, 271, 324, 325
Sovetkin, Dmitry K. 146, 150, 151
Soviet Army 38, 199, 203, 207–209, 216, 217, 294
Soviet Navy 25, 203, 214, 271
Soviet Women's Antifascist Committee 289
space exploration/conquest 16, 27, 29, 46, 47, 55, 84, 112, 131, 149, 156, 157, 160–175
Space Station Solyut, 342
spaceships 25, 156, 160–162, 166, 168, 170, 173, 174, 269, 342
Spasokukotsky, Sergei I. 218, 219, 221
Spontaneous fission 18
sputniks 153, 160, 162, 167, 173
St. Petersburg University →Leningrad University
Stalin, Iosif V. 4, 5, 17, 18, 20, 21, 31, 32, 34, 51, 53, 54, 58–60, 72, 214, 233, 313, 342
Stalin's anti-Semitic campaign 14, 37, 78, 184, 187, 199, 203, 208, 209, 221, 284, 289, 295–297, 349
Stalin's terrors 8, 47, 53, 54, 59, 63, 93, 112, 155, 168, 208, 257, 272, 289, 333, 344
Stalingrad Institute of Mechanics 154
Stalingrad radio center 337
Stalingrad Tractor Plant 154
Starchakov, Aleksandr 223
State Committee of Defense →State Defense Committee
State Defense Committee 16, 72
State Duma 208
State Historical Museum 68
State Optical Research Institute 7
Stebut, Ivan A. 229, 230
Steklov Institute 7, 25–29, 86, 87, 90, 96, 170, 260, 350
Steklov, Vladimir A. 26
Stepanov, Vyacheslav V. 261
Stern, Lina S. 289
Sternfeld, Ari A. 271, 272
Stoletov, Aleksandr G. 78, 89, 90, 171, 245
Stolyarov, G. A. 25
Streletsky, Nikolai S. 280, 332
Struchkov, Yury T. 49–51
Structure of Molecules and the Chemical Bond (Syrkin and Dyatkina) 43
Sudakov, Konstantin V. 182, 183
Suess, Eduard 107
Sukhoi Experimental Construction Bureau 154
Sukhoi, Pavel O. 154, 274, 275
Sumgin, Mikhail I. 116
Surguchev, Mikhail L. 140, 141
Sushkin, Petr P. 98, 325
Suslova, Nadezhda P. 191
Sverzhensky Clinic of the Diseases of the Ear, Throat, and Nose 220
Sverzhensky Institute of Otolaryngology 220
Sverzhensky, Lyudvig I. 220
Syrkin, Yakov K. 43, 44, 136

T

T-34 tank 150
Tager, Iosif L. 295
Talrose, Victor L. 40
Tamm, Igor E. 7–9, 12, 14, 18, 20, 21, 88, 267, 268, 318

Tarakanov, German I. 244
Tareev Clinic of Nephrology, Internal, and Occupational Medicine 194
Tareev, Evgeny M. 194, 195, 296
Tareeva, Irina E. 296
Tatishchev, Vasily N. 113, 114
Technical University of Communication and Informatics 172
teorminimum 34
Terentiev, Boris P. 286
Tereshkova, Valentina V. 148, 166
Terpigorev, Aleksandr M. 302, 303
Thatcher, Margaret 42, 135
theory of resonance 43, 93, 328
thermonuclear reactions 9, 16, 20, 21
Third Clinic of Surgery (Budapest) 184
Third Medical Institute 181
Tikhonov, Andrei N. 86, 87
Tikhonravov, Mikhail K. 272
Tillo, Aleksei A. 125
Timakov, Vladimir D. 197, 198, 296
TIME magazine 200
Timiryazev Academy 47, 57, 122, 133, 225–253, 287, 321, 339
Timiryazev Biological Museum 322
Timiryazev Institute of Biology 288
Timiryazev, Kliment A. 78, 82, 121, 187, 205, 224, 228, 235, 239, 240, 245, 246, 249, 288, 329
Timofeev, Nikolai N. 244
Timofeev-Ressovsky, Nikolai V. 64, 65
Tolstoy, Aleksey 112
Tolstoy, Leo (Lev) N. 191, 242
Tomsk Polytechnic Institute 107, 108, 281–282
Tomsk University 207
Topchiev Institute of Petroleum 95, 134, 285
Topchiev, Aleksandr V. 134, 137, 286
Topchieva, Klavdiya V. 94
Totskoe proving ground 213
Tretyakov Filial 313–315
Trifonov, Yury V. 335
Trisvyatsky, Lev A. 227
Trotsky, Lev D. 54, 216
Troyan, Nadezhda V. 189
Troyekurovskoe Cemetery 348
Tselikov, Aleksandr I. 142, 156
Tsereteli Art Gallery 311
Tsereteli, Zurab K. 309, 310
Tsiolkovsky, Konstantin E. 46, 78, 148, 164, 168, 170, 171, 175, 271
Tsitsin Main Botanical Garden 325, 326
Tsitsin, Nikolai V. 289, 325, 326, 335
Tsvetaeva, Marina I. 310
Tugarinov, Aleksei I. 113
Tulaikov, Nikolai M. 248, 251
Tumansky, Sergei K. 275, 276
Tupolev, Aleksei A. 155
Tupolev, Andrei N. 154, 155, 157, 168, 272, 274, 276
Turbin, Nikolai V. 339, 340
Tursky, Mitrofan K. 242

U
U-2 plane 270
Ukrainian Institute of Physics and Technology 34
Umov, Nikolai A. 88, 90, 262
UNESCO protection 255
United Institute of Nuclear Research 90
University Clinical Hospitals 133, 183, 185, 189
University of Basel 262
University of Berlin 6–7, 241
University of Geneva 289
University of Paris 203
University of Pittsburgh 145
University of Science and Technology (MISiS) 45, 141–143, 282
University of Uzbekistan 62
University of Vienna 107
University of Warsaw →Warsaw University

University of Wisconsin (Madison) 44
University of Yuryev (Yuryev – Dorpat – Tartu) 97, 190, 207, 208, 217
Urazov, Georgy G. 286
US National Academy of Sciences 63, 86, 96
Ushakova, Elizaveta I. 290
Usmanov, Alisher 341
Usov, Mikhail A. 107, 108
Usov, Sergei A. 326
Uspekhi fizicheskikh nauk (*Uspekhi*) 21, 267
Ustinov, Nikolai D. 280
Uvarov, Vladimir V. 151

V
Vadim Sidur Museum 318
Vagankovskoe Cemetery 139, 201, 348
Vainshtein, Boris K. 41, 42, 143
Vainshtein, L. A. 31
Vannikov, B. L. 343
Vargas de Bedemar Museum of Forestry 243
Vargas de Bedemar, Alfons R. 243
Vavilov, Nikolai I. 7, 56–61, 65, 228, 232–234, 246, 248, 250, 251, 267, 309, 310, 340
Vavilov, Petr P. 230, 231
Vavilov, Sergei I. 5–7, 12, 32, 88, 90, 91, 146, 158, 267, 268, 272
Vavilov-Cherenkov Effect →Cherenkov Effect
Veksler, Vladimir I. 11, 88, 158, 268
Vereshchagin, Gleb Yu. 127
Vernadsky Crimean Federal University 17n
Vernadsky Geological Museum 112, 323
Vernadsky Institute 303
Vernadsky, Vladimir I. 18, 69, 78, 85, 93, 100, 106, 109, 111–113, 116, 248, 281, 302, 323
Vernov, Sergei N. 269
Veterinarian Institute 223
Vilyams (Williams), Vasily R. 123, 246–248, 251
Vilyams Fodder Research Institute 247
Vinogradov Clinic of Internal Medicine 183
Vinogradov, Aleksandr P. 303
Vinogradov, Georgy V. 296
Vinogradov, Ivan M. 25, 26, 29, 87, 91, 260
Vinogradov, Vladimir N. 296
Vinogradov, Vladimir V. 296
Vinter, Aleksandr V. 280
Vishnevsky Institute of Surgery 203, 212, 213, 292, 335
Vishnevsky, Aleksandr A. 212, 297
Vishnevsky, Aleksandr V. 212, 297
Voeikov, Aleksandr I. 126
Voevodsky, Vladislav V. 93, 286
Volkov, Vladislav M. 346
Volpin, Mark E. 50
Volsky, Anton N. 286
Vorobiev, Sergei A. 231
Vorobiev, Vladimir P. 297
Voronezh University 12, 217, 302
Voskresenskoe Cemetery (Saratov) 56, 60
Voskresensky, Leonid A. 272
Vostryakovskoe Cemetery 9, 348
Vovsi, Miron S. 208
Vul, Bentsion M. 11, 269
Vvedenskoe Cemetery 211, 349
Vvedensky, Boris A. 280
Vyshelessky, Sergei N. 222, 223

W
Wallenberg, Raoul 39, 39n
Walton, Ernest 319
Warsaw University 98, 127, 324
Warsaw Veterinarian Institute 223
Watson, James D. 341
Welch, William H. 316
Wolf Prize 96

Woodward, Robert B. 36
World Fare (Philadelphia) 132

Y
Yakovlev, Aleksandr S. 148, 276
Yakovlev, Nikolai N. 325
Yakovlev, S. A. 31
Yangel, Mikhail K. 272
Yeltsin, Boris N. 208, 255, 256
Yudin, Sergei S. 203, 297
Yudin, Vadim M. 290
Yurevsky Veterinarian Institute (today, Tartu) 223
Yurov, Ivan A. 208, 209

Z
Zakharkin, Grigory A. 190, 191
Zakharov, Aleksei A. 346
Zakharov, Sergei A. 123
Zavenyagin, Avraami P. 343, 344
Zavoisky, Evgeny K. 24, 345
Zbarsky, Boris I. 297
Zeldovich, Yakov B. 29, 37, 269
Zelinsky Institute of Organic Chemistry 47, 49, 50
Zelinsky, Nikolai D. 48, 49, 69, 70, 94, 281, 285, 339
Zenkevich, Lev A. 303
Zernov, Pavel M. 279, 280
Zhdanov, Andrei A. 296
Zhdanov, Dmitry A. 297, 298
Zhebrak, Anton R. 339, 340
Zhegalov, Sergei I. 227
Zheleznov, Nikolai I. 226, 235
Zhemchugov, Praskov 203
Zhukov, Boris P. 286
Zhukov, Georgy K. 213
Zhukovsky Air Force Academy → Zhukovsky-Gagarin Air Force Academy
Zhukovsky Institute of Aero-hydrodynamics 27, 28, 149, 154, 170
Zhukovsky, Nikolai E. 45, 76, 82, 132, 146–149, 154, 157, 227, 272, 277, 344, 345
Zhukovsky-Gagarin Air Force Academy 148, 162, 260, 276
Zilber, Lev A. 63, 197, 198, 199, 290, 346
Zimin, Anatoly I. 151
Zoo Park 98, 326
Zoological Museum of Moscow University 69, 98
Zoological Museum of the Academy of Sciences 98
Zoological Museum (Petrograd) 127
Zubov, Nikolai N. 128, 303
Zworykin, Vladimir K. 144, 145

Index of Artists and Architects

A
Aizenshtadt, M. B. 124
Aleshin, Nina A. 328
Allakhverdyants, A. S. 118, 210
Andreev, N. A. 70, 211
Anikushin, Mikhail K. 113, 180, 276, 296
Avakyan, V. A. 17
Azgur, Z. I. 216, 292

B
Babichev, A. B. 112
Baburin, M. F. 312
Bakh, Robert 228, 228n, 238
Balashov, A. M. 248
Baranov, L. M. 206, 313
Barshch, M. O. 164
Bazhenov, V. I. 71
Bazhenova, Z. V. 121
Belashova, E. F. 218
Benua, Nikolai L. 248
Bichukov, A. A. 214
Bogushevskaya, N. V. 110
Bondarenko, P. I. 44, 119, 125
Brodsky, B. N. 103, 122
Burganov, A. N. 220
Bykovsky, K. M. 69, 187

C
Chaikov, I. M. 120, 121
Chebotarev, V. S. 117
Chernetsov, N. N. 241
Chernov, Yu. L. 160, 169
Chubarov, E. I. 313

D
Derunov, D. I. 105
Doronina, L. Ya. 122
Dydykin, N. V. 81
Dzyubanov, P. V. 242

F
Faidysh-Krandievsky, A. P. 164, 166, 171
Fokin, S. I. 117
Frangulyan, G. V. 244

G
Galliulin, M. 102
Geondzhian, R. G. 115
Gerlenshtein, E. S. 122
Giliardi, Domenico 69
Glebov, A. K. 89
Gritsuk, M. A. 114
Grubbe, A. V. 111

I
Iofan, Boris M. 73, 333

K
Kampioni, Santino P. 102, 248
Kantor, Maxim K. 316
Kapinus, L. M. 123
Kazakov, Matvei 69, 148
Kazakov, S. V. 121
Kazanskaya, Ekaterina 13
Kenig, P. V. 106, 124
Kerbel, L. E. 116, 157, 166, 187, 189
Khazan, F. S. 272

Khodamaev, N. P. 110
Klimenkova-Krauze, L. K. 127
Klykov, V. M. 330
Kolchin, A. L. 164
Komov, N. I. 48
Konenkov, Sergei T. 104, 108, 192, 212, 301, 316, 317
Korolev, B. D. 107
Korsukov, A. D. 123
Kotov, I. 103
Kovarskaya, E. M. 108
Kovner, S. Ya. 160, 196
Kozlovsky, I. I. 71, 82
Kudrov, V. G. 193
Kuznetsov, N. I. 125
Kvinikhidze, O. V. 121, 245

L
Lavrova, L. N. 115
Lazarev, L. K. 9
Lebedeva, S. D. 123
Lemport, Vladimir 313
Lishev, V. V. 218
Listopad, M. F. 119
Litovchenko, M. L. 81, 126

M
Makhtin, S. O. 247
Manizer, Matvei G. 76, 81, 82, 329
Manashkin, V. S. 127
Manuilov, A. A. 115
Matyushin, L. N. 229, 234, 235
Merkulov, S. D. 71, 313
Miklashevskaya, V. V. 106, 113, 293
Minaev, Vladimir 309
Monihetti, I.A. 160
Mukhina, Vera T. 75, 315
Muravin, L. D. 128

N
Neizvestny, Ernst 256, 266
Neroda, G. V. 81, 114, 148
Neroda, Yu. G. 126
Nikogosyan, Nikolai B. 32, 106, 128, 150, 243, 315, 319
Niss-Goldman, N. I. 119

O
Ognev, G. A. 239
Olenin, M. P. 213
Orekhov, Yury 310
Ostrovskaya, S. L. 128
Ozolina, G. I. 221
Ozolina, T. N. 118

P
Pavlov, L. 328
Pekarev, A. V. 126, 127
Pisarev, N. A. 127
Pisarevsky, L. M. 105
Polyakova, T. R. 125
Popov, S. D. 111
Postol, A. G. 184
Pototsky, G. V. 304, 305
Preobrazhenskaya, E. B. 312
Prisyazhnyuk, A. V. 125

R
Rabin, A. S. 118
Rabinovich, I. A. 82, 122, 125
Rakitina, E. N. 118
Rodionov, Artem 40
Rudko, N. I. 227
Rudnev, Lev 73
Rukavishnikov, A. I. 113, 180
Rukavishnikov, I. M. 16, 75
Ryabichev, Aleksandr 135
Ryabichev, Dmitry 322

S
Samoilova, Natalya K. 328
Sardaryan, Ashot 249
Selikhanov, S. I. 89
Selivanov, Nikolai 22
Selivanov, Vasily 22
Sergeev, A. I. 123

Shakarov, G. A. 245
Shapiro, Peter 309
Shaposhnikov, D. S. 107
Shcherbakov, S. A. 131, 167, 185
Shcherbakov, S. S. 167
Shchipakin, L. 164
Shervud, V. O. 189
Shevnukov, N. S. 191
Shorchev, M. A. 329
Shorchev, L. K. 329
Shults, G. A. 245
Shvarts, D. P. 106
Sidur, Vadim A. 268, 282, 318, 318n
Silis, Nikolai 265
Sogoyan, Mikael 273
Soloviev, S. U. 180
Soskiev, V. B. 170
Stempkovsky, N. M. 116
Stepanyan, A. L. 106
Strakhovskaya, Mariya M. 239
Stritovich, D. A. 200, 204
Suminov, K. S. 56

T
Tazba, L. V. 186
Teneta, A. I. 119
Toidze, G. M. 113
Tomsky, N. V. 66
Tsereteli, Zurab K. 84, 221, 309–311, 314
Tsigal, Vladimir E. 178, 193, 331
Tyurin, E. D. 71

V
Vasiltsov, V. 328
Vatagin, V. A. 321
Velmina, I. A. 103
Velmina, N. A. 116
Ventsel, N. K. 121
Vilensky, E. M. 100, 111
Vilensky, Z. M. 149, 300, 312, 323
Vishkarev, A. P. 123
Vitali, I. P. 146
Volnukhin, Sergei M. 330
Vrubel, A. 294
Vuchetich, E. 305, 315

Y
Yakovleva, I. L. 122
Yanson-Manizer, Elena A. 76
Yarosh, E. S. 126

Z
Zalessky, I. P. 187
Zharenova, L. 328
Zhilov, D. S. 58
Zhilyardi, D. 146
Zhukov, A. F. 337